Learn how the best-run American companies use these
EIGHT BASIC PRINCIPLES
to stay on top of the heap!

One: A bias for action: a preference for doing something—anything—rather than sending a question through cycles and cycles of analyses and committee reports.

Two: Staying close to the customer—learning his preferences and catering to them.

Three: Autonomy and entrepreneurship—breaking the corporation into small companies and encouraging them to think independently and competitively.

Four: Productivity through people—creating in *all* employees the awareness that their best efforts are essential and that they will share in the rewards of the company's success.

Five: Hands-on, value driven—insisting that executives keep in touch with the firm's essential business.

Six: Stick to the knitting—remaining with the business the company knows best.

Seven: Simple form, lean staff—few administrative layers, few people at the upper levels.

Eight: Simultaneous loose-tight properties—fostering a climate where there is dedication to the central values of the company combined with tolerance for all employees who accept those values.

In Search of Excellence

"One of the most useful books to appear in a long while"
—The Atlantic Monthly

"Should be used as a text in every graduate school of business in the country"
—Savvy

IN SEARCH OF
EXCELLENCE

Authors' Note: Excellence 2003

As we look back on the two decades since the publication of *Excellence*, our main feeling is delight. Delight that so many people embraced the book. Delight in that we think we mainly got it right.

Our main detractors point to the decline of some of the companies we featured. They miss the point, which was to learn from those who'd had a long run of success, just as we learn from athletes in their prime. We weren't writing *Forever Excellent*, just as it would be absurd to expect any great athlete not to age. (The sheer longevity of Procter & Gamble's success is, however, intriguing.) And to the most hardheaded of our critics on this dimension: had you bought an "excellence index" when we published and held it through 2002, your total return would have been 1,300 percent, compared to around 800 percent for the Dow and 600 percent for the S&P 500.*

What we take most pleasure in is sticking to our guns in putting several theory chapters up front: "The Rational Model," "Man Waiting for Motivation," and "Managing Ambiguity and Paradox." We told the readers they could skip these. Now we don't think so. They should be read and are just as relevant today as when we originally published.

---

* Over a 5-,10-, or 20-year period an "excellence index," an unweighted basket of the public companies, outperformed both the Dow and S&P 500. Dan Ackman, "Excellence Sought—and Found," Forbes.com, October 10, 2002.

In brief, we were saying the following (but don't skip the chapters). First, people and organizations are not "rational" in the ways strategy, business, and organization are typically taught. It's dangerous to try to force a simplistic and misguided rationality on the way we manage. You cannot just manage "by the numbers." Don't even think about it.

Second, most of the management systems that treat people as "factors of production," as cogs in an industrial machine, are inherently demotivating. People are wonderfully different and complex. Leaders need to set people free to help, not try to harness them.

Third, the world is a confusing place, full of ambiguity. The hardest thing to manage is the "soft stuff," especially culture. Yet without serious attention to the so-called soft stuff, leaders will fail.

Nor would we change our eight attributes of excellence, though we're perfectly aware that another researcher, looking at the same data, might pick a different set. These eight pretty clearly describe what's different about the top performers:

A bias for action. In its simplest terms, this says "get out there and try something." Just as you don't learn anything in science without experimenting, you don't learn anything in business without trying, failing, and trying again. The trick, and it's a tough one, is a common cultural understanding of what kind of failure is okay and what kind leads to disaster. But don't kid yourself. No amount of analysis, especially market research, will lead to true innovation.

Close to the customer. This may be the hardest to accomplish and perhaps where our sample of excellent companies—IBM, Hewlett-Packard, K Mart, and even McDonald's—got off the track. It's hard. There's so much to pay attention to inside an organization that has time to understand customers, especially when the set of customers includes distributors and wonderfully irrational end users. Yet Procter's skill at keeping everyone in the organization in close touch with customers, combined with a formidable innovative capability, may explain that company's incredibly long history of success.

Autonomy and entrepreneurship. Even if you're big, act small. Organizations are simply collections of people, and people don't relate well to big, abstract entities. If you want to understand the success of Johnson & Johnson, 3M, Wal-Mart, and the original HP, look to the fact that they organize themselves into small, relatively independent units, held together by common goals and cultural norms.

Productivity through people. As the youngsters say, "Duh!" What else counts in an organization except people? Everyone gives lip service to the importance of their people, yet only a few really treat them as other than cannon fodder. One of the best examples we've ever seen was Delta Airlines with its "family feeling," which was so special that in 1982 employees banded together to spend a total of $30 million in payroll deductions to give their employer its first Boeing 767, the *Spirit of Delta*. Sadly, Delta lost that family feeling, maybe when it merged with Western Airlines.

Hands-on, value-driven. The idea is simple. Figure out what your company should stand for, what would give your people the most pride. Then actively manage toward that value system. Remember that profit is to business as breathing is to life. The top companies make meaning, not just money.

Stick to the knitting. Except for one or two notable exceptions, for example, Warren Buffett's Berkshire Hathaway and Jack Welch's GE, business diversity almost never works. Be particularly leery of the word *synergy*, which sounds great—who doesn't want 1 + 1 to equal 3? Well, our observation then and now is that big mergers rarely work. Further, nothing screws up a successful business more than hyperfast growth.

Simple form, lean staff. Though organizations are inherently quite complicated, one ought not make them more so via complex organizational arrangements. Install a simple and workable structure; people will figure out the rest. Keep staff to a minimum, outsource a lot of staff activities, or use time-limited, project-oriented task forces (another form of the line organization). Big staffs, and most career staff people, always seem to get in the way of the folks in organizations who get the real work done.

Simultaneous loose-tight properties. Sorry about this chapter heading, but it does say what we mean. Any well-functioning organization is neither centralized nor decentralized but a wonderful combination of both. Around most dimensions the best companies, then and now, are loose. They give people exceptional freedom to do things their own way.† At the same time, the great companies are highly centralized around a few crucial dimensions: the central values that make up their culture, one or two (no more) top strategic priorities, and a few key financial indicators.

Those are our eight attributes, then and now. Both of us have done a great deal of writing since *Excellence*, and we've expressed what we've seen in different terms. But we've never done better than in this book. The attributes are just that: attributes, not principles. But until something clearly better comes along, we'll stick with these.

—Tom Peters

Bob Waterman

November 2003

---

† This seems to be true in every setting. We remember talking recently with FedEx delivery people, whose work lives seemed, to us, to be terribly regimented. They liked their jobs and what they said they liked most was "freedom to do things their own way."

IN SEARCH OF
EXCELLENCE

Lessons from America's Best-Run Companies

by Thomas J. Peters
and Robert H. Waterman, Jr.

HarperBusiness Essentials
A HarperBusiness Book
An Imprint of HarperCollins*Publishers*

IN SEARCH OF EXCELLENCE. Copyright © 2004, 1982 by Thomas J. Peters and Robert H. Waterman, Jr. All rights reserved. Printed in the United States of America. No part of this book may be used or reproduced in any manner whatsoever without written permission except in the case of brief quotations embodied in critical articles and reviews. For information address HarperCollins Publishers Inc., 10 East 53rd Street, New York, NY 10022.

HarperCollins books may be purchased for educational, business, or sales promotional use. For information please write: Special Markets Department, HarperCollins Publishers Inc., 10 East 53rd Street, New York, NY 10022.

First HarperBusiness Essentials edition published 2004

The Library of Congress has catalogued the previous edition as follows:

Peters, Thomas J.
    In search of excellence.

    Includes bibliographical references and index.
    1. Industrial management—United States.   I. Waterman,
Robert H.   II. Title.
HD70.U5P424   1983   658'.00973   83-12442
0-446-38507-7 (U.S.A.)

ISBN 0-06-054878-9

04 05 06 07 08 RRD 10 9 8 7 6 5 4 3 2 1

For Gene Webb and Lew Young, who inspired the book.
And for Judy, Robb, and Kendall, who are a
source of continuing inspiration.

Contents

Acknowledgments

Two people made this book eminently more readable than it otherwise might have been—John Cox and Jennifer Futernick. John took our early, much too long, far too redundant draft and helped us blast our way through the quite substantial barrier between unwieldy early scribblings and something that resembled a book. John was also enormously helpful in tightening our manuscript during the throes of final drafting. Jennifer Futernick was the other major contributor to the writing process. We had originally involved Jennifer as research librarian, someone to help us get the facts straight. As it turns out, however, Jennifer also has an unusual sense for what works on paper and what doesn't. She not only was incredibly helpful with the detailed line editing and checking of our manuscript, but of far more importance, she repeatedly called our attention to structural problems, statements that we really couldn't support with the facts, and redundancy. Jennifer adopted the book as her own and put extraordinary hours and matchless care into its development.

McKinsey and Company as a whole was gracious in support of the time we devoted to the excellent company research. Our conclusions have been tempered and sharpened by many of our partners' contributions. A few deserve special thanks. Warren Cannon and Ron Daniel have been believers—once lonely believers—since the inception of this research. Jon Katzenbach has relentlessly cheered us on from the start of our effort. Allan Kennedy acted as an intellectual prod and allowed us to test some of our flakier ideas out in the heat of battle, while others blanched. Herb Henzler, from Munich, championed our early efforts, believed in their pragmatic value, and helped us test that belief in client settings.

Also, Julien Phillips, Don Gogel, Jim Bennett, Jim Balloun, Rajat Gupta, Bill Price, Ron Bancroft, David Meen, and Bill Matassoni at McKinsey have been active "promoters" and refiners of the excellent-company material.

Our intellectual debt to four brilliant thinkers about organization effectiveness is especially keen. Karl Weick at Cornell, Gene Webb and Hal Leavitt at Stanford, and Herb Simon at Carnegie-Mellon have thumbed their noses at conventional thinking for decades. The first three have been rich personal sources of inspiration. We have benefited, as have so many others, from Professor Simon's powerful notions, although via the written word only.

Obviously, the most important contributors to the excellent-company research are our friends in the companies themselves. Three stand out. Rene McPherson, from Dana (and now Stanford), is a source of unparalleled inspiration. His record as Dana's chairman suggests that mortals *can* move mountains. John Young of Hewlett-Packard offered time and, more important, vital encouragement when we most needed it—at the beginning. Tait Elder of 3M (and now Allied Corporation) taught us more about innovation than we thought knowable.

Other notable contributors include Stan Little of Boeing, Stan Abramson of Westinghouse, Allan Gilbert of Emerson, Jim Shapiro and Ken Stahl of Xerox, Larry Small and Jack Heilshorn of Citibank, Jack Welch of GE, and Buck Rodgers of IBM. Their belief that we had something worthwhile to say was even more vital than the facts they provided.

Of equal importance are the hundreds of nameless participants in the over two hundred groups to whom we have presented the material. So many have added yet another Digital or IBM story—confirming, denying, and always sharpening our arguments.

Among those nameless, many of our Stanford Graduate School

of Business students are included. In the book, we are tough on the business schools; but we are angered at the faculties, not the students. The students care a great deal about the quality of American management. Gary Bello, a Stanford Sloan Program student from General Motors, shaped more of our approach to the "Productivity Through People" chapter than he will ever know; his brand of caring was inspiring.

Drs. Max Günther and Hermann Grabherr at Siemens, in West Germany, also deserve special mention. They were intrigued by our early research and were active supporters of our survey. Moreover, their relentless and always thoughtful questioning was often critical to the honing of our ideas.

A book like this is not only the product of current research and colleagues. Our predispositions are the product of a lifetime. In that vein, Tom owes special thanks to his mother, Evelyn Peters, who inculcated the restless curiosity that led to this research, and to vital early mentors, especially Dick Anderson, Blake van Leer, and Walter Minnick. Bob owes special thanks to his mother, Virginia Waterman, who shaped his early notions of excellence, and to his father, Robert Waterman, who taught him the values of initiative and integrity through personal example.

We can do little more than bow down in gratitude before the people to whom we dedicate the book   Gene, Lew, and Judy. Gene Webb, at Stanford, has been a source of total support to Tom for nearly fifteen years. Lew Young, at *Business Week,* above all personally cares about the ideas (and this book, if it is anything, is about caring and commitment). Judy Waterman was Bob's original teacher on the importance of enthusiasm and "nonrational" approaches to life.

Of critical importance was the contribution of those who worked exceedingly long hours to clarify early drafts, to search for facts and to type the multiple drafts for the manuscript and the speeches that led to the manuscript. These include Janet Collier, Nancy Kaible, Nancy Rynd, Patty Bulena, and Sylvia Osterman. We also gladly acknowledge the very special contribution of Kay Dann, who not only helped type, but most important—acting as the authors' administrative assistant—maintained calm in the midst of our often frenetic, out-of-control activity.

The last word of acknowledgment delightedly goes to Robbin Reynolds at Harper & Row. She found us, nurtured our early incoherent work, railed at us when we (often) needed it, and patted us on the back (often undeservingly) to keep us going. Thanks, Robbin, for the faith.

Special Acknowledgment: David G. Anderson

We want to pay special tribute to David G. Anderson, who was and remains a very special contributor to the excellent-company research. David, who was at McKinsey at the time and who is now working toward his Ph.D. at Stanford University, while on leave from McKinsey, worked on the project from its inception. He organized and personally conducted many of the field interviews and did the original financial research on the excellent-company sample. Most important, David fully contributed his awesome intellectual prowess toward originating and shaping many of the core ideas that have emerged over the years. Further, his forceful responses to emerging ideas (pro and con) usually represented the acid test of intellectual merit. For example, David pushed the central role of champions and what we call limited autonomy positions. And he turned our attention to the literature on the power of perceived control. His commitment that arose out of dogged intellectual pursuit of key ideas was and is a true source of inspiration—and comfort.

In addition, David has been a frequent speaker and leader of feedback sessions as the work has progressed. Finally, David was a prime contributor to the tenth chapter of this book, "Stick to the Knitting."

Preface

There are a few observations that may help the reader through the pages ahead. We collected the data on which this book is based and distilled them into eight basic findings. Some readers may say that the findings are motherhoods, but that's not true. Each finding in and of itself may seem a platitude (close to the customer, productivity through people), but the intensity of the way in which the excellent companies execute the eight—especially when compared with their competitors—is as rare as a smog-free day in Los Angeles.

Second, we hazard that Chapters 3 and 4 may be daunting, because they are devoted largely to theory. They can be skipped (or read last), but we *do* suggest that the reader skim them, at least, and consider giving them careful attention. We urge this, because the eight basics of management excellence don't just "work because they work." They work because they make exceptional sense. The deepest needs of hundreds of thousands of individuals are tapped—exploited, if you will—by the excellent companies, and their success reflects, sometimes without their knowing it, a sound theoretical basis. Moreover, we think readers may be pleasantly surprised to see how interesting the theory is. It is not, we would add, new or untested; most of the theory has stood the scientific test of time and defied refutation. It merely has been ignored, by and large, by managers and management writers.

We also would like to say here that the majority of the excellent companies are not McKinsey clients. McKinsey supported the research and the writing but did not influence our selection of companies.

Introduction

We had decided, after dinner, to spend a second night in Washington. Our business day had taken us beyond the last convenient flight out. We had no hotel reservations, but were near the new Four Seasons, had stayed there once before, and liked it. As we walked through the lobby wondering how best to plead our case for a room, we braced for the usual chilly shoulder accorded to late-comers. To our astonishment the concierge looked up, smiled, called us by name, and asked how we were. She remembered our names! We knew in a flash why in the space of a brief year the Four Seasons had become the "place to stay" in the District and was a rare first-year holder of the venerated four-star rating.

Good for them, you are thinking, but why the big deal? Well, the incident hit us with some force because for the past several years we have been studying corporate excellence. For us, one of the main clues to corporate excellence has come to be just such incidents of unusual effort on the part of apparently ordinary employees. When we found not one but a host of such incidents, we were pretty certain we were on the track of an exceptional situation. What's more, we were fairly sure we would find sustained financial performance that was as exceptional as the employees' performance.

Other images come to mind. We were in another Washington, the state this time, talking to a group of Boeing executives about our research and making the point that excellent companies seem to take all sorts of special trouble to foster, nourish, and care for what we call "product champions"—those individuals who believe so strongly in their ideas that they take it on themselves to damn the bureaucracy and maneuver their projects through the system and out to the customer. Someone piped up: "Champions! Our problem is we can't kill them." Then Bob Withington, who was present when it all happened, went on to tell the story about how Boeing had *really* won the contracts for the swept-wing B-47, which was later to become the highly successful first commercial jet, the 707. He also told the story about how Boeing *really* won the contract for the B-52, which was to have been a turboprop design until Boeing was able to demonstrate the advantages of B-52 as jet aircraft.

For us, the fascination of the first story was the saga of a little band of Boeing engineers poring through German files on the day Nazi labs were occupied by the Allied forces. In so doing, they quickly confirmed their own ideas on the enormous advantages of swept-wing design. Then it was the drama halfway around the world in Seattle of the subsequent rush to verify swept-wing design in the wind tunnel and the surprising finding that if the engine couldn't be on the aircraft body, it was best suspended out in front of the wing. The second story told of one long, sleepless weekend in a Dayton hotel where a small team of engineers completely redesigned the B-52, wrote and produced a 33-page bound proposal, and presented it to the Air Force just seventy-two hours later, the following Monday. (This tiny team of champions, moreover, presented the proposal complete with a finely sculpted scale model, which they had made out of balsa and other materials purchased during the weekend for $15 at a local hobby shop.) These were both fine tales of little teams of people going to extraordinary lengths to get results on behalf of a truly unusual corporation. Yet the Boeing pattern emerged as the norm at companies as disparate as 3M and IBM; small, competitive bands of pragmatic bureaucracy-beaters, the source of much innovation.

To cite yet another example, we dropped by a small calculator and electronics store the other day to buy a programmable calculator. The salesman's product knowledge, enthusiasm, and interest in us were striking and naturally we were inquisitive. As it happened, he was not a store employee at all, but a twenty-eight-year-old Hewlett-Packard (HP) development engineer getting some first-hand experience in the users' response to the HP product line. We had heard that HP prides itself on its closeness to the customer and that a typical assignment for a new MBA or electrical engineer was to get involved in a job that included the practical aspects of product introduction. Damn! Here was an HP engineer behaving as enthusiastically as any salesman you'd ever want to see.

Wherever we have been in the world, from Australia to Europe to Japan, we can't help but be impressed by the high standard of cleanliness and consistency of service we find in every McDonald's hamburger outlet. Not everyone likes the product, nor the concept of McDonald's as a worldwide expression of American culture, but it really *is* extraordinary to find the kind of quality assurance McDonald's has achieved worldwide in a service business. (Controlling quality in a service business is a particularly difficult problem. Unlike manufacturing, in which one can sample what comes off the line and reject bad lots, what gets produced in service businesses and what gets consumed happens at the same time and in the same place. One must ensure that tens of thousands of people throughout the company are adhering roughly to the same high standard and that they all understand the company's conception of and genuine concern for quality.)

We recalled a conversation that took place one sunny, calm spring day in a canoe on the mirror waters of Lake Geneva, years before this research was undertaken. One of us was teaching at IMEDE, a business school in Lausanne, and was visiting an old colleague. His ventures had had him traveling constantly, which distressed his wife, so he up and started a chain of McDonald's outlets in Switzerland, which kept him home but left his wife, who was born in Geneva, in a state of xenophobic shock. (She got over it as soon as the Swiss became loyal McDonald's customers.) He was talking about his early impressions of McDonald's, and commented, "You know, one of the things that strike me most about McDonald's is their people orientation. During the seven years I was at McKinsey, I never saw a client that seemed to care so much about its people."

Another friend described for us why, in a recent major computer system purchase for a hospital, he chose International Business Machines. "Many of the others were ahead of IBM in technological wizardry," he noted. "And heaven knows their software is easier to use. But IBM alone took the trouble to get to know us. They interviewed extensively up and down the line. They talked our language, no mumbo jumbo on computer innards. Their price was fully twenty-five percent higher. But they provided unparalleled guarantees of reliability and service. They even went so far as to arrange a backup connection with a local steel company in case our system crashed. Their presentations were to the point. Everything about them smacked of assurance and success. Our decision, even with severe budget pressure, was really easy."

We hear stories every other day about the Japanese companies, their unique culture and their proclivity for meeting, singing company songs, and chanting the corporate litany. Now, that sort of thing is usually dismissed as not relevant in America, because who among us can imagine such tribal behavior in U.S. companies? But American examples do exist. For anyone who has not seen it, it is hard to imagine the hoopla and excitement that attend the weekly Monday night Rally of people who sell plastic bowls—Tupperware bowls. Similar goings on at Mary Kay Cosmetics were the subject of a segment done by Morley Safer on *Sixty Minutes*. Those examples might be dismissed as peculiar to selling a certain kind of product. On the other hand, at HP, the regular beer bust for all hands is a normal part of each division's approach to keeping everyone in touch. And one of us went through an IBM sales training program early in his career; we sang songs every morning and got just as enthusiastic (well, almost as enthusiastic) as the workers in a Japanese company.

In teaching workshops for clients or students, we often use a case built around Delta Airlines' unique management style. We who travel a lot are apt to tell a story or two about the material assistance we have gotten from Delta's gate employees while scrambling to make a last-minute connection. The last time we did it, one executive raised his hand and said, "Now, let me tell you how it really is at Delta." As we were preparing for what was clearly to be a challenge to our thesis, the individual went on to describe a story of exceptional service from Delta that made ours pale by comparison. His wife had inadvertently missed out on a super saver ticket because the family had moved and, owing to a technicality, the ticket price was no longer valid. She called to complain. Delta's president intervened personally and, being there at the time, met her at the gate to give her the new ticket.

Anyone who has been in brand management at Procter & Gamble sincerely believes that P&G is successful more for its unusual commitment to product quality than for its legendary marketing prowess. One of our favorite images is that of a P&G executive, red in the face, furiously asserting to a class in a Stanford summer executive program that P&G "does too make the best toilet paper on the market, and just because the product is toilet paper, or soap for that matter, doesn't mean that P&G doesn't make it a damn sight better than anyone else." (As in most of the excellent companies, these basic values run deep. P&G once refused to substitute an inferior ingredient in its soap, even though it meant not meeting the Army's pressing needs during the war—the Civil War.)

Finally, at Frito-Lay we hear stories, perhaps apocryphal, probably not—it doesn't matter—about people slogging through sleet, mud, hail, snow, and rain. They are not delivering the mail. They are potato chip salesmen, upholding the "99.5% service level" * in which the entire Frito organization takes such pride—and which is the source of its unparalleled success.

And the stories go on. What really fascinated us as we began to pursue our survey of corporate excellence was that the more we dug, the more we realized the excellent companies abounded in such stories and imagery. We began to realize that these companies

---

* At Frito, a mom and pop store in Missoula, Montana, or the flagship *Safeway* in Oakland, California, each stands the same 99.5 percent chance of getting a daily call from its Frito route salesman.

had cultures as strong as any Japanese organization. And the trappings of cultural excellence seemed recognizable, no matter what the industry. Whatever the business, by and large the companies were doing the same, sometimes cornball, always intense, always repetitive things to make sure all employees were buying into their culture—or opting out.

Moreover, to our initial surprise, the content of the culture was invariably limited to just a handful of themes. Whether bending tin, frying hamburgers, or providing rooms for rent, virtually all of the excellent companies had, it seemed, defined themselves as de facto service businesses. Customers reign supreme. They are not treated to untested technology or unnecessary goldplating. They are the recipients of products that last, service delivered promptly.

Quality and service, then, were invariable hallmarks. To get them, of course, everyone's cooperation is required, not just mighty labors from the top 200. The excellent companies require and demand extraordinary performance from the average man. (Dana's former chairman, Rene McPherson, says that neither the few destructive laggards nor the handful of brilliant performers are the key. Instead, he urges attention to the care, feeding, and unshackling of the average man.) We labeled it "productivity through people." All companies pay it lip service. Few deliver.

Finally, it dawned on us that we did not have to look all the way to Japan for models with which to attack the corporate malaise that has us in its vicelike grip. We have a host of big American companies that are doing it right from the standpoint of all their constituents—customers, employees, shareholders, and the public at large. They've been doing it right for years. We have simply not paid enough attention to their example. Nor have we attempted to analyze the degree to which what they instinctively do is fully consistent with sound theory.

Discussions of management psychology have long focused on theory X or theory Y, the value of job enrichment, and, now, quality circles. These don't go far toward explaining the magic of the turned-on work force in Japan or in the American excellent company, but useful theory does exist. The psychologist Ernest Becker, for example, has staked out a major supporting theoretical position, albeit one ignored by most management analysts. He argues that man is driven by an essential "dualism"; he needs both to be a part of something and to stick out. He needs at one and the same time to be a conforming member of a winning team and to be a star in his own right.

About the winning team, Becker notes: "Society . . . is a vehicle for earthly heroism. . . . Man transcends death by finding meaning for his life. . . . It is the burning desire for the creature to count. . . . What man really fears is not so much extinction, but extinction with *insignificance* . . . Ritual is the technique for giving life. His sense of self worth is constituted symbolically, his cherished narcissism feeds on symbols, on an abstract idea of his own worth. [Man's] natural yearning can be fed limitlessly in the domain of symbols." He adds: *"Men fashion unfreedom* [a large measure of conformity] *as a bribe for self-perpetuation."* In other words, men willingly shackle themselves to the nine-to-five if only the cause is perceived to be in some sense great. The company can actually provide the same resonance as does the exclusive club or honorary society.

At the same time, however, each of us needs to stick out—even, or maybe particularly, in the winning institution. So we observed, time and again, extraordinary energy exerted above and beyond the call of duty when the worker (shop floor worker, sales assistant, desk clerk) is given even a modicum of apparent control over his or her destiny. An experiment in psychology consistent with this major field of inquiry underscores the point. Adult subjects were given some complex puzzles to solve and a proofreading chore. In the background was a loud, randomly occurring distracting noise; to be specific, it was "a combination of two people speaking Spanish, one speaking Armenian, a mimeograph machine running, a desk calculator, and a typewriter, and street noise—producing a composite, nondistinguishable roar." The subjects were split into two groups. Individuals in one set were just told to work at the task. Individuals in the other were provided with a button to push to turn off the noise, "a modern analog of control—the off switch." The group with the off switch solved five times the number of puzzles as their cohorts and made but a tiny fraction of the number of proofreading errors. Now for the kicker: "... none of the subjects in the off switch group ever used the switch. The mere knowledge that one can exert control made the difference."

The best-managed companies, and a few others, act in accordance with these theories. For example, the manager of a 100-person sales branch rented the Meadowlands Stadium (New Jersey) for the evening. After work, his salesmen ran onto the stadium's field through the players' tunnel. As each emerged, the electronic scoreboard beamed his name to the assembled crowd. Executives from corporate headquarters, employees from other offices, and family and friends were present, cheering loudly.

The company is IBM. With one act (most nonexcellent companies would write it off as too corny, too lavish, or both), IBM simultaneously reaffirmed its heroic dimension (satisfying the individual's need to be a part of something great) and its concern for individual self-expression (the need to stick out). IBM is bridging an apparent paradox. If there is one striking feature of the excellent companies, it is this ability to manage ambiguity and paradox. What our rational economist friends tell us ought not to be possible the excellent companies do routinely.

Frito's chips and Maytag's washers ought to be commodities; a 99.5 percent service level for mom and pop stores is silly—until you look at the margins, until you see the market share. The problem in America is that our fascination with the tools of management obscures our apparent ignorance of the art. Our tools are biased toward measurement and analysis. We can measure the costs. But with these tools alone we can't really elaborate on the value of a turned-on Maytag or Caterpillar work force churning out quality products or a Frito-Lay salesperson going that extra mile for the ordinary customer.

Worse, our tools force us into a rational bent that views askance the very sources of innovation in the excellent companies: irrational product champions at 3M, product-line proliferation and duplication at Digital Equipment Corporation, the intense internal competition among P&G brand managers. Alfred Sloan successfully introduced overlap at General Motors in the 1920s; extensive and purposeful overlap has existed among IBM divisions' product lines to spur internal competition for almost as long. But few rationalists seem to buy it, even today. They don't like overlap; they do like tidiness. They don't like mistakes; they do like meticulous planning. They don't like not knowing what everyone is up to; they do like controls. They build big staffs. Meanwhile, Wang Labs or 3M or Bloomingdale's is ten new product introductions and months ahead.

So we take some exception to traditional theory, principally because our evidence about how human beings work—individually and in large groups—leads us to revise several important economic tenets dealing with size (scale economies), precision (limits to analysis), and the ability to achieve extraordinary results (particularly quality) with quite average people.

The findings from the excellent companies amount to an upbeat message. There is good news from America. Good management practice today is not resident only in Japan. But, more important, the good news comes from treating people decently and asking them to shine, and from producing things that work. Scale efficiencies give way to small units with turned-on people. Precisely planned R&D efforts aimed at big bang products are replaced by armies of dedicated champions. A numbing focus on cost gives way to an enhancing focus on quality. Hierarchy and three-piece suits give way to first names, shirtsleeves, hoopla, and project-based flexibility. Working according to fat rule books is replaced by everyone's contributing.

Even management's job becomes more fun. Instead of brain games in the sterile ivory tower, it's shaping values and reinforcing through coaching and evangelism in the field—with the worker and in support of the cherished product.

This book will elaborate more fully on what we have just described. It will define what we mean by excellence. It is an attempt to generalize about what the excellent companies seem to be doing that the rest are not, and to buttress our observations on the excellent companies with sound social and economic theory. And, finally, it will employ field data too often overlooked in books on management—namely, specific, concrete examples from the companies themselves.

PART ONE

THE SAVING REMNANT

1

Successful American Companies

The Belgian Surrealist René Magritte painted a series of pipes and entitled the series *Ceci n'est pas une pipe (This is not a pipe)*. The picture of the thing is not the thing. In the same way, an organization chart is not a company, nor a new strategy an automatic answer to corporate grief. We all know this; but like as not, when trouble lurks, we call for a new strategy and probably reorganize. And when we reorganize, we usually stop at rearranging the boxes on the chart. The odds are high that nothing much will change. We will have chaos, even useful chaos for a while, but eventually the old culture will prevail. Old habit patterns persist.

At a gut level, all of us know that much more goes into the process of keeping a large organization vital and responsive than the policy statements, new strategies, plans, budgets, and organization charts can possibly depict. But all too often we behave as though we don't know it. If we want change, we fiddle with the strategy. Or we change the structure. Perhaps the time has come to change our ways.

Early in 1977, a general concern with the problems of management effectiveness, and a particular concern with the nature of the relationship between strategy, structure, and management effectiveness, led us to assemble two internal task forces at McKinsey & Company. One was to review our thinking on strategy, and the other was to go back to the drawing board on organizational effectiveness. It was, if you like, McKinsey's version of applied research. We (the authors) were the leaders of the project on organizational effectiveness.

A natural first step was to talk extensively to executives around the world who were known for their skill, experience, and wisdom on the question of organizational design. We found that they, too, shared our disquiet about conventional approaches. All were uncomfortable with the limitations of the usual structural solutions, especially the latest aberration, the complex matrix form. Yet they were skeptical about the usefulness of any known tools, doubting they were up to the task of revitalizing and redirecting billion-dollar giants.

In fact, the most helpful ideas were coming from the strangest places. Way back in 1962, the business historian Alfred Chandler wrote *Strategy and Structure,* in which he expressed the very powerful notion that structure follows strategy. And the conventional wisdom in 1977, when we started our work, was that Chandler's dictum had the makings of universal truth. Get the strategic plan down on paper and the right organization structure will pop out with ease, grace, and beauty. Chandler's idea *was* important, no doubt about that; but when Chandler conceived it everyone was diversifying, and what Chandler most clearly captured was that a strategy of broad diversification dictates a structure marked by decentralization. Form follows function. For the period following World War II through about 1970, Chandler's advice was enough to cause (or maintain) a revolution in management practice that was directionally correct.

But as we explored the subject, we found that strategy rarely seemed to dictate unique structural solutions. Moreover, the crucial problems in strategy were most often those of execution and continuous adaptation: getting it done, staying flexible. And that to a very large extent meant going far beyond strategy to issues of organizing—structure, people, and the like. So the problem of management effectiveness threatened to prove distressingly circular. The dearth of practical additions to old ways of thought was painfully apparent. It was never so clear as in 1980, when U.S. managers, beset by obvious problems of stagnation, leaped to adopt Japanese management practices, ignoring the cultural difference, so much wider than even the vast expanse of the Pacific would suggest.

Our next step in 1977 was to look beyond practicing businessmen for help. We visited a dozen business schools in the United States and Europe (Japan doesn't have business schools). The theorists from academe, we found, were wrestling with the same concerns. Our timing was good. The state of theory is in refreshing disarray, but moving toward a new consensus; some few researchers continue to write about structure, particularly that latest and most modish variant, the matrix. But primarily the ferment is around another stream of thoughts that follows from some startling ideas about the limited capacity of decision makers to handle information and reach what we usually think of as "rational" decisions, and the even lesser likelihood that large collectives (i.e., organizations) will automatically execute the complex strategic design of the rationalists.

The stream that today's researchers are tapping is an old one, started in the late 1930s by Elton Mayo and Chester Barnard, both at Harvard. In various ways, both challenged ideas put forward by Max Weber, who defined the bureaucratic form of organization, and Frederick Taylor, who implied that management really can be made into an exact science. Weber had pooh-poohed charismatic leadership and doted on bureaucracy; its rule-driven, impersonal form, hc said, was the only way to assure long-term survival. Taylor, of course, is the source of the time and motion approach to efficiency: if only you can divide work up into enough discrete, wholly programmed pieces and then put the pieces back together in a truly optimum way, why then you'll have a truly top-performing unit.

Mayo started out four-square in the mainstream of the rationalist school and ended up challenging, de facto, a good bit of it. On the shop floors of Western Electric's Hawthorne plant, he tried to demonstrate that better work place hygiene would have a direct and positive effect on worker productivity. So he turned up the lights. Productivity went up, as predicted. Then, as he prepared to turn his attention to another factor, he routinely turned the lights back down. Productivity went up again! For us, the very important message of the research that these actions spawned, and a theme we shall return to continually in the book, is that it is *attention to employees,* not work conditions per se, that has the dominant impact on productivity. (Many of our best companies, one friend observed, seem to reduce management to merely creating "an endless stream of Hawthorne effects.") It doesn't fit the rationalist view.

Chester Barnard, speaking from the chief executive's perspective (he had been president of New Jersey Bell), asserted that a leader's role is to harness the social forces in the organization, to shape and guide values. He described good managers as value shapers concerned with the informal social properties of organization. He contrasted them with mere manipulators of formal rewards and systems, who dealt only with the narrower concept of short-term efficiency.

Barnard's concepts, although quickly picked up by Herbert Simon (who subsequently won a Nobel prize for his efforts), lay otherwise dormant for thirty years while the primary management disputes focused on structure attendant to postwar growth, the burning issue of the era.

But then, as the first wave of decentralizing structure proved less than a panacea for all time and its successor, the matrix, ran into continuous troubles born of complexity, Barnard and Simon's ideas triggered a new wave of thinking. On the theory side, the exemplars were Karl Weick of Cornell and James March of Stanford, who attacked the rational model with a vengeance.

Weick suggests that organizations learn and adapt v-e-r-y slowly. They pay obsessive attention to habitual internal cues, long after their practical value has lost all meaning. Important strategic business assumptions (e.g., a control versus a risk-taking bias) are buried deep in the minutiae of management systems and other habitual routines whose origins have long been obscured by time. Our favorite example of the point was provided by a friend who early in his career was receiving instruction as a bank teller. One operation involved hand-sorting 80-column punched cards, and the woman teaching him could do it as fast as lightning. "Bzzzzzzt" went the deck of cards in her hands, and they were all sorted and neatly stacked. Our friend was all thumbs.

"How long have you been doing this?" he asked her.

"About ten years," she estimated.

"Well," said he, anxious to learn, "what's that operation for?"

"To tell you the truth"—Bzzzzzzt, another deck sorted—"I really don't know."

Weick supposes that the inflexibility stems from the mechanical pictures of organizations we carry in our heads; he says, for instance: "Chronic use of the military metaphor leads people repeatedly to overlook a different kind of organization, one that values improvisation rather than forecasting, dwells on opportunities rather than constraints, discovers new actions rather than defends past actions, values arguments more highly than serenity and encourages doubt and contradiction rather than belief."

March goes even further than Weick. He has introduced, only slightly facetiously, the garbage can as organizational metaphor. March pictures the way organizations learn and make decisions as streams of problems, solutions, participants, and choice opportunities interacting almost randomly to carry the organization toward the future. His observations about large organizations recall President Truman's wry prophecy about the vexations lying in wait for his successor, as recounted by Richard E. Neustadt. "He'll sit here," Truman would remark (tapping his desk for emphasis), "and he'll say, 'Do this! Do that!' And nothing will happen. Poor Ike—it won't be a bit like the army. He'll find it very frustrating."

Other researchers have recently begun to accumulate data that support these unconventional views. The researcher Henry Mintzberg, of Canada's McGill University made one of the few rigorous studies of how effective managers use their time. They don't regularly block out large chunks of time for planning, organizing, motivating, and controlling, as most authorities suggest they ought. Their time, on the contrary, is fragmented, the average interval devoted to any one issue being *nine minutes*. Andrew Pettigrew, a British researcher, studied the politics of strategic decision making and was fascinated by the inertial properties of organizations. He showed that companies often hold on to flagrantly faulty assumptions about their world for as long as a decade, despite overwhelming evidence that that world has changed and they probably should too. (A wealth of recent examples of what Pettigrew had in mind is provided by the several American industries currently undergoing deregulation—airlines, trucking, banks, savings and loans, telecommunications.)

Among our early contacts were managers from long-term top-performing companies: IBM, 3M, Procter & Gamble, Delta Airlines. As we reflected on the new school of theoretical thinking, it began to dawn on us that the intangibles that those managers described were much more consistent with Weick and March than with Taylor or Chandler. We heard talk of organizational cultures, the family feeling, small is beautiful, simplicity rather than complexity, hoopla associated with quality products. In short, we found the obvious, that the individual human being still counts. Building up organizations that take note of his or her limits (e.g., information-processing ability) and strengths (e.g., the power flowing from commitment and enthusiasm) was their bread and butter.

CRITERIA FOR SUCCESS

For the first two years we worked mainly on the problem of expanding our diagnostic and remedial kit beyond the traditional tools for business problem solving, which then concentrated on strategy and structural approaches.

Indeed, many friends outside our task force felt that we should simply take a new look at the structural question in organizing. As decentralization had been the wave of the fifties and sixties, they said, and the so-called matrix the modish but quite obviously ineffective structure of the seventies, what then would be the structural form of the eighties? We chose to go another route. As important as the structural issues undoubtedly are, we quickly concluded that they are only a small part of the total issue of management effectiveness. The very word "organizing," for instance, begs the question, "Organize for what?" For the large corporations we were interested in, the answer to that question was almost always to build some sort of major new corporate capability—that is, to become more innovative, to be better marketers, to permanently improve labor relations, or to build some other skill which that corporation did not then possess.

An excellent example is McDonald's. As successful as that corporation was in the United States, doing well abroad meant more than creating an international division. In the case of McDonald's it meant, among other things, teaching the German public what a hamburger is. To become less dependent on government sales, Boeing had to build the skill to sell its wares in the commercial marketplace, a feat most of its competitors never could pull off. Such skill building, adding new muscle, shucking old habits, getting really good at something new to the culture, is difficult. That sort of thing clearly goes beyond structure.

So we needed more to work with than new ideas on structure. A good clue to what we were up to is contained in a remark by Fletcher Byrom, chairman and chief executive of Koppers: "I think an inflexible organization chart which assumes that anyone in a given position will perform exactly the same way his predecessor did, is ridiculous. He won't. Therefore, the organization ought to shift and adjust and adapt to the fact that there's a new person in the spot." There is no such thing as a good structural answer apart from people considerations, and vice versa. We went further. Our research told us that any intelligent approach to organizing had to encompass, and treat as interdependent, at least seven variables: structure, strategy, people, management style, systems and procedures, guiding concepts and shared values (i.e., culture), and the present and hoped-for corporate strengths or skills. We defined this idea more precisely and elaborated what came to be known as the McKinsey 7-S Framework (see figure on next page). With a bit of stretching, cutting, and fitting, we made all seven variables start with the letter S and invented a logo to go with it. Anthony Athos at the Harvard Business School gave us the courage to do it that way, urging that without the memory hooks provided by alliteration, our stuff was just too hard to explain, too easily forgettable.

Hokey as the alliteration first seemed, four years' experience throughout the world has borne out our hunch that the framework would help immeasurably in forcing explicit thought about not only

McKINSEY 7-S FRAMEWORK ©

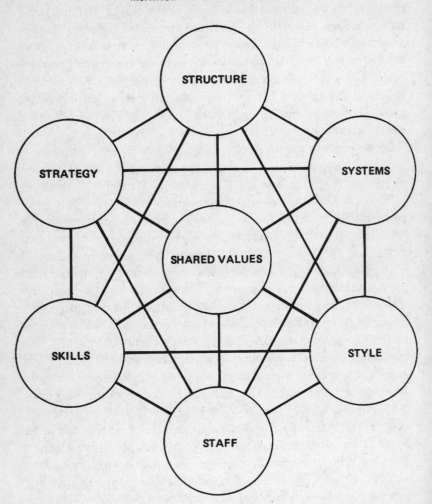

the hardware—strategy and structure—but also about the software of organization—style, systems, staff (people), skills, and shared values. The framework, which some of our waggish colleagues have come to call the happy atom, seems to have caught on around the world as a useful way to think about organizing.* Richard Pascale and Anthony Athos, who assisted us in our concept development, used it as the conceptual underpinning for *The Art of Japanese Management*. Harvey Wagner, a friend at the University of North Carolina and an eminent scholar in the hard-nosed field of decision sciences, uses the model to teach business policy. He said recently, "You guys have taken all the mystery out of my class. They [his students] use the framework and all the issues in the case pop right to the surface."

In retrospect, what our framework has really done is to remind the world of professional managers that "soft is hard." It has enabled us to say, in effect, "All that stuff you have been dismissing for so long as the intractable, irrational, intuitive, informal organization *can* be managed. Clearly, it has as much or more to do with the way things work (or don't) around your companies as the formal structures and strategies do. Not only are you foolish to ignore it, but here's a way to think about it. Here are some tools for managing it. Here, really, is the way to develop a new skill."

But there was still something missing. True, we had expanded our diagnostic tool kit by quantum steps. True, we had observed managers apparently getting more done because they could *pay attention* with seven S's instead of just two. True, by recognizing that real change in large institutions is a function of at least seven hunks of complexity, we were made appropriately more humble about the difficulty of changing a large institution in any fundamental way. But, at the same time, we were short on practical design ideas,

* We were hardly the first to invent a multi-variable framework. Harold Leavitt's "Leavitt's Diamond," for instance (task, structure, people, information and control, environment), has now influenced generations of managers. We were fortunate in enjoying good timing. Managers beset with seemingly intractable problems and years of frustration with strategy and structure shifts were finally ready for a new view by 1980. Moreover, putting the stamp of McKinsey, long known for its hard-nosed approach to management problem solving, behind the new model added immense power.

especially for the "soft S's." Building new corporate capability wasn't the simple converse of describing and understanding what's not working, just as designing a good bridge takes more than understanding why some bridges fail. We now had far better mental equipment for pinpointing the cause of organizational malaise, which was good, and we had enhanced our ability to determine what was working despite the structure and ought to be left alone, which was even better. But we needed to enrich our "vocabulary" of design patterns and ideas.

Accordingly, we decided to take a look at management excellence itself. We had put that item on the agenda early in our project, but the real impetus came when the managing directors of Royal Dutch/Shell Group asked us to help them with a one-day seminar on innovation. To fit what we had to offer with Shell's request, we chose a double meaning for the word "innovation." In addition to what might normally be thought of—creative people developing marketable new products and services—we added a twist that is central to our concern with change in big institutions. We asserted that innovative companies not only are unusually good at producing commercially viable new widgets; *innovative companies are especially adroit at continually responding to change of any sort in their environments.* Unlike Andrew Pettigrew's inertial organizations, when the environment changes, these companies change too. As the needs of their customers shift, the skills of their competitors improve, the mood of the public perturbates, the forces of international trade realign, and government regulations shift, these companies tack, revamp, adjust, transform, and adapt. In short, as a whole culture, they innovate.

That concept of innovation seemed to us to define the task of the truly excellent manager or management team. The companies that seemed to us to have achieved that kind of innovative performance were the ones we labeled excellent companies.

We gave our presentation to Royal Dutch/Shell Group on July 4, 1979, and if this research has a birthday, that was it. What fascinated us even more than the effort in The Netherlands, however, was the reaction we subsequently got from a few companies like HP and 3M that we had contacted in preparation for our discussions with Shell. They were intrigued with the subject we were pursuing and urged us on.

Largely because of that, several months later we put together a team and undertook a full-blown project on the subject of excellence as we had defined it—continuously innovative big companies. This was mainly funded by McKinsey, with some support from interested clients. At that point we chose seventy-five highly regarded companies, and in the winter of 1979–80 conducted intense, structured interviews in about half these organizations. The remainder we initially studied through secondary channels, principally press coverage and annual reports for the last twenty-five years; we have since conducted intensive interviews with more than twenty of those companies. (We also studied some underachieving companies for purposes of comparison, but we didn't concentrate much on this, as we felt we had plenty of insight into underachievement through our combined twenty-four years in the management consulting business.)

Our findings were a pleasant surprise. The project showed, more clearly than could have been hoped for, that the excellent companies were, above all, brilliant on the basics. Tools didn't substitute for thinking. Intellect didn't overpower wisdom. Analysis didn't impede action. Rather, these companies worked hard to keep things simple in a complex world. They persisted. They insisted on top quality. They fawned on their customers. They listened to their employees and treated them like adults. They allowed their innovative product and service "champions" long tethers. They allowed some chaos in return for quick action and regular experimentation.

The eight attributes that emerged to characterize most nearly the distinction of the excellent, innovative companies go as follows:

1. *A bias for action,* for getting on with it. Even though these companies may be analytical in their approach to decision making, they are not paralyzed by that fact (as so many others seem to be). In many of these companies the standard operating procedure is "Do it, fix it, try it." Says a Digital Equipment Corporation senior executive, for example, "When we've got a big problem here, we

grab ten senior guys and stick them in a room for a week. They come up with an answer *and* implement it." Moreover, the companies are experimenters supreme. Instead of allowing 250 engineers and marketers to work on a new product in isolation for fifteen months, they form bands of 5 to 25 and test ideas out on a customer, often with inexpensive prototypes, within a matter of weeks. What is striking is the host of practical devices the excellent companies employ, to maintain corporate fleetness of foot and counter the stultification that almost inevitably comes with size.

2. *Close to the customer*. These companies learn from the people they serve. They provide unparalleled quality, service, and reliability—things that work and last. They succeed in differentiating—*à la* Frito-Lay (potato chips), Maytag (washers), or Tupperware—the most commodity-like products. IBM's marketing vice president, Francis G. (Buck) Rodgers, says, "It's a shame that, in so many companies, whenever you get good service, it's an exception." Not so at the excellent companies. Everyone gets into the act. Many of the innovative companies got their best product ideas from customers. That comes from listening, intently and regularly.

3. *Autonomy and entrepreneurship*. The innovative companies foster many leaders and many innovators throughout the organization. They are a hive of what we've come to call champions; 3M has been described as "so intent on innovation that its essential atmosphere seems not like that of a large corporation but rather a loose network of laboratories and cubbyholes populated by feverish inventors and dauntless entrepreneurs who let their imaginations fly in all directions." They don't try to hold everyone on so short a rein that he can't be creative. They encourage practical risk taking, and support good tries. They follow Fletcher Byrom's ninth commandment: "Make sure you generate a reasonable number of mistakes."

4. *Productivity through people*. The excellent companies treat the rank and file as the root source of quality and productivity gain. They do not foster we/they labor attitudes or regard capital investment as the fundamental source of efficiency improvement. As Thomas J. Watson, Jr., said of his company, "IBM's philosophy is largely contained in three simple beliefs. I want to begin with what

I think is the most important: *our respect for the individual.* This is a simple concept, but in IBM it occupies a major portion of management time." Texas Instruments' chairman Mark Shepherd talks about it in terms of every worker being "seen as a source of ideas, not just acting as a pair of hands"; each of his more than *9,000* People Involvement Program, or PIP, teams (TI's quality circles) does contribute to the company's sparkling productivity record.

5. *Hands-on, value driven.* Thomas Watson, Jr., said that "the basic philosophy of an organization has far more to do with its achievements than do technological or economic resources, organizational structure, innovation and timing." Watson and HP's William Hewlett are legendary for walking the plant floors. McDonald's Ray Kroc regularly visits stores and assesses them on the factors the company holds dear, Q.S.C. & V. (Quality, Service, Cleanliness, and Value).

6. *Stick to the knitting.* Robert W. Johnson, former Johnson & Johnson chairman, put it this way: "Never acquire a business you don't know how to run." Or as Edward G. Harness, past chief executive at Procter & Gamble, said, "This company has never left its base. We seek to be anything but a conglomerate." While there were a few exceptions, the odds for excellent performance seem strongly to favor those companies that stay reasonably close to businesses they know.

7. *Simple form, lean staff.* As big as most of the companies we have looked at are, none when we looked at it was formally run with a matrix organization structure, and some which had tried that form had abandoned it. The underlying structural forms and systems in the excellent companies are elegantly simple. Top-level staffs are lean; it is not uncommon to find a corporate staff of fewer than 100 people running multi-billion-dollar enterprises.

8. *Simultaneous loose-tight properties.* The excellent companies are both centralized and decentralized. For the most part, as we have said, they have pushed autonomy down to the shop floor or product development team. On the other hand, they are fanatic centralists around the few core values they hold dear. 3M is marked by barely organized chaos surrounding its product champions. Yet

one analyst argues, "The brainwashed members of an extremist political sect are no more conformist in their central beliefs." At Digital the chaos is so rampant that one executive noted, "Damn few
people know who they work for." Yet Digital's fetish for reliability
is more rigidly adhered to than any outsider could imagine.

Most of these eight attributes are not startling. Some, if not
most, are "motherhoods." But as Rene McPherson says, "Almost
everybody agrees, 'people are our most important asset.' Yet almost
none really lives it." The excellent companies live their commitment
to people, as they also do their preference for action—any action—
over countless standing committees and endless 500-page studies,
their fetish about quality and service standards that others, using
optimization techniques, would consider pipe dreams; and their insistence on regular initiative (practical autonomy) from tens of
thousands, not just 200 designated $75,000-a-year thinkers.

Above all, the *intensity itself,* stemming from strongly held beliefs, marks these companies. During our first round of interviews,
we could "feel it." The language used in talking about people was
different. The expectation of regular contributions was different.
The love of the product and customer was palpable. And we felt
different ourselves, walking around an HP or 3M facility watching
groups at work and play, from the way we had in most of the more
bureaucratic institutions we have had experience with. It was
watching busy bands of engineers, salesmen, and manufacturers casually hammering out problems in a conference room in St. Paul in
February; even a customer was there. It was seeing an HP division
manager's office ($100 million unit), tiny, wall-less, on the factory
floor, shared with a secretary. It was seeing Dana's new chairman,
Gerald Mitchell, bearhugging a colleague in the hall after lunch in
the Toledo headquarters. It was very far removed from silent board
rooms marked by dim lights, somber presentations, rows of staffers
lined up along the walls with calculators glowing, and the endless
click of the slide projector as analysis after analysis lit up the
screen.

We should note that not all eight attributes were present or conspicuous to the same degree in all of the excellent companies we

studied. But in every case at least a preponderance of the eight was clearly visible, quite distinctive. We believe, moreover, that the eight are conspicuously absent in most large companies today. Or if they are not absent, they are so well disguised you'd hardly notice them, let alone pick them out as distinguishing traits. Far too many managers have lost sight of the basics, in our opinion: quick action, service to customers, practical innovation, and the fact that you can't get any of these without virtually everyone's commitment.

So, on the one hand, the traits are obvious. Presenting the material to students who have no business experience can lead to yawns. "The customer comes first, second, third," we say. "Doesn't *everyone* know that?" is the implied (or actual) response. On the other hand, seasoned audiences usually react with enthusiasm. They know that this material is important, that Buck Rodgers was right when he said good service is the exception. And they are heartened that the "magic" of a P&G and IBM is simply getting the basics right, not possessing twenty more IQ points per man or woman. (We sometimes urge them not to be so heartened. The process of acquiring or sharpening the basics to anything like the excellent companies' obsessive level, after all, is a lot harder than coming up with a "strategic breakthrough" in one's head.)

American companies are being stymied not only by their staffs (about which more hereafter), but also by their structures and systems, both of which inhibit action. One of our favorite examples is shown in a diagram drawn by a manager of a would-be new venture in a moderately high technology business (figure on next page).

The circles in this diagram represent organizational units—for example, the one containing MSD is the Management Sciences Division—and the straight lines depict the formal linkages (standing committees) that are involved in launching a new product. There are 223 such formal linkages. Needless to say, the company is hardly first to the marketplace with any new product. The irony, and the tragedy, is that each of the 223 linkages taken by itself makes perfectly good sense. Well-meaning, rational people designed each link for a reason that made sense at the time—for example, a committee was formed to ensure that a glitch between sales and mar-

NEW PRODUCT SIGN-OFF

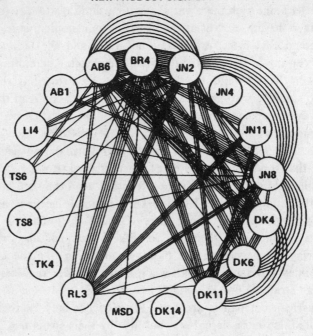

keting, arising in the last product rollout, is not repeated. The trouble is that the total picture as it inexorably emerged, amusing as it might be to a C. Northcote Parkinson, captures action like a fly in a spider's web and drains the life out of it. The other sad fact is that when we use this diagram in presentations, we don't draw shouts of "Absurd." Instead we draw sighs, nervous laughter, and the occasional volunteer who says, "If you really want a humdinger, you should map our process."

THE RESEARCH

The sample of sixty-two companies* was never intended to be perfectly representative of U.S. industry as a whole, although we think we have captured a fairly broad spectrum. Nor did we try to be too precise at the beginning about what we meant by excellence or innovation. We were afraid at that point that had we tried to be too precise we would lose the essence of what we thought we were after, as in E. B. White's account of humor, which "can be dissected, as a frog, but the thing dies in the process and the innards are discouraging to any but the pure scientific mind." What we really wanted and got with our original group was a list of companies considered to be innovative and excellent by an informed group of observers of the business scene—businessmen, consultants, members of the business press, and business academics. The companies were grouped into various categories to ensure that we would have enough representation in the industry segments we were interested in. (See table on next page.) The industry categories include but are not limited to:

1. High-technology companies, such as Digital Equipment, Hewlett-Packard (HP), Intel, and Texas Instruments (TI)

2. Consumer goods companies, such as Procter & Gamble (P&G), Chesebrough-Pond's, and Johnson & Johnson (J&J)

3. General industrial goods companies of interest (a catch-all ob-

---

* There were seventy-five in the original sample. Thirteen were European. These were dropped from the analysis because they do not represent a fair cross-section of European companies.

EXCELLENT COMPANY SURVEY

STRUCTURED INTERVIEWS PLUS 25-YEAR LITERATURE REVIEW		
High technology	**Consumer goods**	**General industrial**
Allen-Bradley †	Blue Bell	Caterpillar Tractor*
Amdahl*	Eastman Kodak*	Dana Corporation*
Digital Equipment*	Frito-Lay (PepsiCo) †	Ingersoll-Rand
Emerson Electric*	General Foods	McDermott
Gould	Johnson & Johnson*	Minnesota Mining &
Hewlett-Packard*	Procter & Gamble*	Manufacturing*
International Business Machines*		
NCR		
Rockwell		
Schlumberger*		
Texas Instruments*		
United Technologies		
Western Electric		
Westinghouse		
Xerox		

LIMITED INTERVIEWS PLUS 25-YEAR LITERATURE REVIEW		
Data General*	Atari (Warner Communications) †	General Motors
General Electric	Avon*	
Hughes Aircraft †	Bristol-Myers*	
Intel*	Chesebrough-Pond's*	
Lockheed	Levi Strauss*	
National Semiconductor*	Mars †	
Raychem*	Maytag*	
TRW	Merck*	
Wang Labs*	Polaroid	
	Revlon*	
	Tupperware (Dart & Kraft) †	

* Passes all hurdles for "excellent" performance, 1961-1980
† Privately held or subsidiary; no extensive public data available,
   but estimated to pass all hurdles for "excellent" performance

Service	Project management	Resource based
Delta Airlines*	Bechtel †	Exxon
Marriott*	Boeing*	
McDonald's*	Fluor*	

American Airlines		Arco
Disney Productions*		Dow Chemical*
K mart*		Du Pont*
Wal-Mart*		Standard Oil (Indiana)/
		Amoco*

viously), which included Caterpillar, Dana, and 3M (Minnesota Mining and Manufacturing)

4. Service companies such as Delta Airlines, Marriott, McDonald's, and Disney Productions

5. Project management companies such as Bechtel and Fluor

6. Resource-based companies such as Atlantic-Richfield (Arco), Dow Chemical, and Exxon.

Conspicuously missing from the list were certain industries which later will be the subject of further study. Although our experience with large financial service institutions, and in particular banks, is extensive, they were thought to be too highly regulated and protected (then) to be of interest. Most chemical and drug companies were left out simply because we didn't get around to them. Finally, we didn't look extensively at small companies; our major concern was and is with how big companies stay alive, well, and innovative. Therefore, few firms in our sample had annual sales of less than $1 billion or histories shorter than twenty years.

As a next-to-last step in choosing the companies to be studied in some depth, we reasoned that no matter what prestige these companies had in the eyes of the rest of the business world, the companies were not truly excellent unless their financial performance supported their halo of esteem. Consequently, we chose and imposed six measures of long-term superiority. Three are measures of growth and long-term wealth creation over a twenty-year period. Three are measures of return on capital and sales. The six are:

1. Compound asset growth from 1961 through 1980 (a "least squares" measure that fits a curve to annual growth data).

2. Compound equity growth from 1961 through 1980 (a "least squares" measure of annual growth data).

3. The average ratio of market value to book value. "Market to book" is a standard approximation for what the economists call "wealth creation" (market value: closing share price times common shares outstanding, divided by common book-value equity as of December 31, 1961 through 1980).

4. Average return on total capital, 1961 through 1980 (net in-

come divided by total invested capital, where total invested capital consists of long-term debt, nonredeemable preferred stock, common equity, and minority interests).

5. Average return on equity, 1961 through 1980.
6. Average return on sales, 1961 through 1980.

In order to qualify as a top performer, a company must have been in the top half of its industry in at least four out of six of these measures over the full twenty-year period (in fact, of the thirty-six companies that qualified, seventeen ranked in the top half on all six measures, and another six ranked in the top half of five of six measures).* Thus, any top performer must have scored well, over the long haul, on both growth measures and absolute measures of economic health.

As a last screen, we applied a measure of innovativeness per se. We asked selected industry experts (e.g., businessmen from within the industry) to rate the companies' twenty-year record of innovation, defined as a continuous flow of industry bellwether products and services and general rapidness of response to changing markets or other external dynamics.

Imposing these criteria meant that nineteen companies dropped from our original list of sixty-two. Of the remaining forty-three, we interviewed twenty-one in depth.† We conducted less extensive interviews in each of the remaining twenty-two. We also conducted extensive interviews at twelve companies that we had put in a "?" category; these were ones that did not pass all the screens but had just barely missed. We also followed all sixty-two closely in the literature for the twenty-five years preceding the study.

Finally, we culled the sample in another fashion. Although we prefer to back up our conclusions with hard evidence from specific companies, we do occasionally say, "They do thus and such." In

---

* "Industries" are the six categories noted previously (e.g., high-technology companies). The comparison base for each industry is a random and statistically valid sample from that industry's total population among *Fortune* 500 companies.

†The set of forty-three includes the thirty-six mentioned above plus seven privately held companies (e.g., Mars) or subsidiaries (e.g., Frito-Lay) that we *estimate* to have passed our financial hurdles but for which verification is not wholly possible because of the absence of public data.

this sense, "They" is a group of exemplars which, without benefit of specific selection criteria, do seem to represent especially well both sound performance and the eight traits we have identified. They are: Bechtel, Boeing, Caterpillar Tractor, Dana, Delta Airlines, Digital Equipment, Emerson Electric, Fluor, Hewlett-Packard, IBM, Johnson & Johnson, McDonald's, Procter & Gamble, and 3M. On the surface, they have little in common; there is no universality of product line. Three are in high technology, one is in packaged goods, one principally makes medical products, two are service businesses, two are involved in project management, and five are basic industrial manufacturers. But each is a hands-on operator, not a holding company or a conglomerate. And while not every plan succeeds, in the day-to-day pursuit of their businesses these companies succeed far more often than they fail.

When we finished our interviews and research, we began to sift and codify our results. It was then, roughly six months after we had started, that we reached the conclusions which are the backbone of this book. We still had a few nagging problems, however. We had used the 7-S framework as the basic structuring device for our interviews and hence chose the same framework as a way of communicating our conclusions, with the result that, at the time, we identified twenty-two attributes of excellence. The whole thing was just too confusing and we were in danger of adding to the complexity railed at in the first place. When that was forcefully pointed out to us by several of the early consumers of our research, we went back to work, and tried to distill the essence of what we were saying in a simpler way. The result, with no material loss to the message, is the eight attributes of excellence we describe.

Several questions always come up when we are discussing our findings. First, people often challenge a few of the companies we have used on the basis of their own personal acquaintance. All big companies have their warts and blemishes; as excellent as we claim some of these companies to be, they are not without fault, and they have made plenty of well-publicized mistakes. Also, one man's excellent company is another's stock market disaster. We don't pretend to account for the perfidy of the market or the whims of investors. The companies *have* performed well over long periods, and that is good enough for us.

Second, we are asked how we know that the companies we have defined as culturally innovative will stay that way. The answer is we don't. GM looked excellent at the time and has since had serious troubles. But it will likely survive those troubles better than the rest of the American auto industry. And again, it performed so well for so long that one cannot help being impressed. So we feel about many of the excellent companies.

Third, why have we added (as the reader will soon see) examples from companies that were not on the original list, and examples from companies that do not fit our original definition of excellence? The reason is that our inquiry into corporate innovation and excellence is a continuing effort and much work has been done since 1979. For example, another team within McKinsey did a special study of excellence in the American consumer goods industry; and yet another has recently completed a study of excellent companies in Canada. A group is hard at work on the question of excellence in the medium-sized—or threshold—companies, the "so far, so good" category. Also, as the original team continues to follow up, we find more reinforcement of the early findings and more examples.

The process has been more powerful than we ever dared imagine. Since the original publication of our findings in *Business Week* in July 1980, we have given over 200 speeches, conducted more than 50 workshops—and spent a lot of time on planes. It is a rare day when we don't run into alumni (or active members) of our survey companies. At Memorex, one of us recently ran into a man who had worked directly, and for years, with Watson, Sr., at IBM. Our list of friends and acquaintances from P&G's brand-management program and IBM's sales program is as long as your arm. An acquaintance from our 3M interviews stays in touch: we have spent several long days with him talking about innovation. The corroboration at times becomes amazingly fine-grained. For instance, we laud HP's informality. Yet one of our colleagues, analyzing highly successful Tandem (founded by ex-HPers) argues that "Tandem's traditional Friday beer busts are more exuberant than HP's." We continue to learn more and more—to add confirmation and modification at a level of detail that greatly reinforces our confidence in the findings.

Finally, we are asked, what about evolution and change? How did these companies get the way they are? Is it always a case of a strong leader at the helm? We must admit that our bias at the beginning was to discount the role of leadership heavily, if for no other reason than that everybody's answer to what's wrong (or right) with whatever organization is its leader. Our strong belief was that the excellent companies had gotten to be the way they are because of a unique set of cultural attributes that distinguish them from the rest, and if we understood those attributes well enough, we could do more than just mutter "leadership" in response to questions like "Why is J&J so good?" Unfortunately, what we found was that associated with almost every excellent company was a strong leader (or two) who seemed to have had a lot to do with making the company excellent in the first place. Many of these companies—for instance, IBM, P&G, Emerson, J&J, and Dana—seem to have taken on their basic character under the tutelage of a very special person. Moreover, they did it at a fairly early stage of their development.

But there is a caveat or two. The excellent companies seem to have developed cultures that have incorporated the values and practices of the great leaders and thus those shared values can be seen to survive for decades after the passing of the original guru. Second, going back to where we started with Chester Barnard, it appears that the real role of the chief executive is to manage the *values* of the organization. We hope that what follows, then, will illuminate just what values ought to be shaped and managed, and that we will thereby have helped to solve the leadership dilemma after all.

PART TWO

TOWARD NEW THEORY

2

The Rational Model

Professionalism in management is regularly equated with hard-headed rationality. We saw it surface at ITT in Harold Geneen's search for the "unshakable facts." It flourished in Vietnam, where success was measured by body counts. Its wizards were the Ford Motor Company's whiz kids, and its grand panjandrum was Robert McNamara. The numerative, rationalist approach to management dominates the business schools. It teaches us that well-trained professional managers can manage anything. It seeks detached, analytical justification for all decisions. It is right enough to be dangerously wrong, and it has arguably led us seriously astray.

It doesn't tell us what the excellent companies have apparently learned. It doesn't teach us to love the customers. It doesn't instruct our leaders in the rock-bottom importance of making the average Joe a hero and a consistent winner. It doesn't show how strongly workers can identify with the work they do if we give them a little say-so. It doesn't tell us why self-generated quality control is so much more effective than inspector-generated quality control. It doesn't tell us to nourish product champions like the first buds in springtime. It doesn't impel us to allow—even encourage, as P&G does—in-house product-line competition, duplication, and even product-to-product cannibalization. It doesn't command that we overspend on quality, overkill on customer service, and make products that last and work. It doesn't show, as Anthony Athos puts it, that "good managers make meanings for people, as well as money."

The rational approach to management misses a lot.

When the two of us went to business school, the strongest department was finance, a majority of the students had engineering degrees (including ourselves), courses in quantitative methods flourished, and the only facts that many of us considered "real data" were the ones we could put numbers on. Those were the *old* days, but the situation hasn't changed much. At least when we went to graduate business school in the 1960s a few students could slip through the system with relative distinction on their innate talents as fine bullshitters. Now they approach class at their peril if they haven't "run the numbers" (translation: done some kind—any kind—of quantitative analysis). Many graduate business students so dread the prospect of a calculator battery failing during the final exam that they take spares, spare batteries, an extra calculator, or both. The word "strategy," which used to mean a damn good idea for knocking the socks off the competition, has often come to be synonymous with the quantitative breakthrough, the analytic coup, market share numbers, learning curve theory, positioning business on a 4- or 9- or 24-box matrix (the matrix idea, straight from mathematics) and putting all of it on a computer.

There are nevertheless faint signs of hope. Courses in strategy are starting to recognize and address the problem of implementation. Courses in manufacturing policy (although overwhelmingly quantitative) are at least edging back into the curriculum. But the "technical jocks," as an ex-plant manager colleague of ours calls them, are still a dominant force in American business thinking. Finance departments are still as strong as ever in the business schools. Talented teachers and gifted students in sales management and manufacturing—the core disciplines of most businesses—are still as scarce (and refreshing) as rain in the desert.

Don't misunderstand us. We are not against quantitative analysis per se. The best of the consumer marketers, such as P&G, Chesebrough-Pond's and Ore-Ida, do crisp to-the-point analysis that is the envy and bedevilment of their competitors. Actually, the companies that we have called excellent are among the best at getting the numbers, analyzing them, and solving problems with them.

Show us a company without a good fact base—a good quantitative picture of its customers, markets, and competitors—and we will show you one in which priorities are set with the most byzantine of political maneuvering.

What we are against is wrong-headed analysis, analysis that is too complex to be useful and too unwieldy to be flexible, analysis that strives to be precise (especially at the wrong time) about the inherently unknowable—such as detailed market forecasts when end use of a new product is still hazy (remember, most early estimates supposed that the market for computers was 50 to 100 units)—and especially analysis done *to* line operators *by* control-oriented, hands-off staffs. TI's Patrick Haggerty insisted that "those who *implement* the plans must *make* the plans"; his renowned strategic planning system was overseen by only three staffers, all temporary, all ex-line officers headed that way again.

We are also against situations in which action stops while planning takes over, the all-too-frequently observed "paralysis through analysis" syndrome. We have watched too many line managers who simply want to get on with their job but are deflated by central staffs that can always find a way to "prove" something won't work, although they have no way of quantifying why it might work. The central staff plays it safe by taking the negative view; and as it gains power, it stamps all verve, life, and initiative out of the company.

Above all, we deplore the unfortunate abuse of the term "rational." Rational means sensible, logical, reasonable, a conclusion flowing from a correct statement of the problem. But rational has come to have a very narrow definition in business analysis. It is the "right" answer, but it's missing all of that messy human stuff, such as good strategies that do not allow for persistent old habits, implementation barriers, and simple human inconsistencies. Take economies of scale. If maximum process efficiency *could* be reached, if *all* suppliers produced flawless supplies and produced them on time, if absenteeism *were* absent, and *if* sloppy human interaction didn't get in the way, then big plants *would* outproduce small ones. But, as the researcher John Child points out in a rare quantification of part of the problem, whereas unionized shops with 10 to 25 employees average 15 lost time days from labor disputes per 1,000 employees per year, facilities with 1,000 or more employees lost on average 2,000 days, or a multiple of 133. Take also innovation. A researcher concluded recently that research effectiveness was inversely related to group size: assemble more than seven people and research effectiveness goes down. Our stories of ten-person "skunk works" out-inventing groups of several hundred are corroborative.

We also resist those who argue that all that stuff (small team zeal, disputes that arise as a function of size alone) is the province of the "art" factor in management. Yes, quantification of these sorts of factors is difficult, probably not even useful. But the factors can certainly be considered sensibly, logically, and fairly precisely in the face of modestly well documented past experience. Is it merely art that leads Motorola president John Mitchell, a tough-minded engineer, to say that he won't allow plants to run much above 1,000, principally "because something just seems to go wrong when you get more people under one roof"? Or is it just an enlightened version of sound reasoning, based on fairly precise recollection of experience? We'd bet on the latter.

Why then, you may ask, was the narrow definition of rationality, the "machines-without-damnable-human-operators" view, apparently adequate for so long? Why was it equal to the task of churning out unparalleled productivity gains, especially after World War II? In part, things were simpler then: the pent-up demand for products after World War II, the absence of tough international competitors, a post-depression work force that felt lucky to have a job at all, and the "high" of being an American worker turning out the best and the brightest of tailfins for a tailfin-hungry world were all factors.

There's another critical reason, too. The management techniques of the last twenty-five years have actually been necessary. As we've said, we're advocates of sound analysis. The best companies on our list combine a tablespoon of sound analysis with a pint of love for the hamburger bun; both are indispensable. Before the rise of the analytic model, the seat-of-the-pants technique was all there was.

And it was wholly inadequate for dealing with a complex world. True, learning to segment markets, to factor in the time value of money, and to do sound cash-flow projection have long since become vital steps to business survival. The trouble arose when those techniques became the pint and love of product became the tablespoon. The analytic tools are there to assist—and they can do so admirably—but they still can't make or sell products.

Whatever the reasons, the United States *was* dominant, and, as George Gilder put it in *Wealth and Poverty,* "the secular rationalist mythology" prevailed. This was so manifestly true that Steve Lohr in a recent *New York Times Magazine* cover story noted that just a decade ago the world feared being bowled over by American management technique, not just our labs, our factories, or even our sheer size. "These American invaders were superior, in [French editor Jean-Jacques] Servan-Schreiber's view, not because of their money resources, or technology but because of their corporate organizational ability—and the genius behind it all was the American corporate manager."

But something has happened in the thirteen years since Servan-Schreiber first published *The American Challenge.* American business has gotten mired in a swamp of economic and political woes, most prominently OPEC and increasing domestic regulation. In truth, however, these problems are shared by many other countries, some of which are now the islands of good news. The performance of many Japanese and West German companies is oft-cited evidence that "it can be done." And, of course, they are more hard-hit by OPEC than we are. Also, they more than we are performing in regulated economies. The German managers, far more than the American managers, must deal continually with the labor unions. And the Japanese and German use of individual economic incentives is, relatively speaking, much weaker than our own. The economist Lester Thurow notes:

Nor have [the United States'] competitors unleashed work effort and savings by increasing income differentials. Indeed, they have done exactly the opposite. If you look at the earnings gap between the top and bottom 10 percent of the population, the West Germans work hard for 36 percent less inequality than we, and the Japanese work even harder with 50 percent less inequality. If income differentials encourage individual initiative, we should be full of initiative, since among industrialized countries, only the French surpass us in terms of inequality.

In *The American Challenge,* Servan-Schreiber suggested that once—recently—we valued our management talent more highly than our technical wizardry. But where, interestingly, does Steve Lohr's quote of Servan-Schreiber appear? Its context is an article entitled "Overhauling America's Business Management," an outright attack on American management skills. Lohr delivers the following broadside: "How quickly things change. Today when foreign executives speak of their American counterparts, they are apt to be more scornful than awestruck, and indeed, the United States appears to be strewn with evidence of managerial failure."

Within the space of a few weeks in late 1980, *Newsweek, Time, The Atlantic Monthly, Dun's Review* (twice), and even *Esquire* carried cover stories on the general theme that the managers were to blame for the sad state of American business—not OPEC, not regulation, not monetary incentives, not even our puny investment expenditures. *Fortune* reported an executive vice president at Honda as saying:

The amount of money [the U.S. auto companies] are spending really doesn't bother me. Please don't misunderstand. The United States is the most technologically advanced country, and the most affluent one. But capital investment alone will not make the difference. In any country, the quality of products and the productivity of workers depend on management. When Detroit changes its management system, we'll see more powerful American competitors.*

---

* The first wave of attack seemed to focus on the beleaguered auto industry, but by mid-1981 it was clear that mature industries were not the only ones in trouble. The Japanese garnered a 70 percent share of the 64K RAM chip market, arguably the bellwether of industrial high technology. Most observers admitted (secretly, if not publicly) that the reason was quality, pure and simple, not investment concentration.

Only a few weeks later, *Fortune* followed up the Honda feature with an article entitled "Europe Outgrows American Management Style," which attacked our shortsightedness, our tendency to shuffle managers instead of building stable institutions, and our lack of care for the products we make.

The complaints against American management seem to fall into five main categories: (1) the business schools are doing us in; (2) the so-called professional managers lack the right perspective; (3) managers don't personally identify with what their companies do; (4) managers don't take enough interest in their people; and (5) top managers and their staff have become isolated in their analytic ivory towers.

The salvo against the business schools seems to have generated the most smoke, for the apparent reason that they symbolize the rest and are easy to criticize. H. Edward Wrapp, the highly regarded business policy professor at the University of Chicago, suggests: "We have created a monster. A colleague noted, and I agree, that the business schools have done more to insure the success of the Japanese and West German invasion of America than any one thing I can think of." Wrapp goes on to deplore the business schools' overemphasis on quantitative methods, a complaint echoed repeatedly in our own research. Steve Lohr apparently agrees, concluding in his *New York Times* article that there is now "a widely held view that the MBA might be part of the current problem." Another critic offered a simple prescription for curing the problem—one with which we don't entirely disagree. Says Michael Thomas, former successful investment banker and, of late, inspired author: "[They] lack liberal arts literacy . . . need a broader vision, a sense of history, perspectives from literature and art . . . I'd close every one of the graduate schools of business. . . ." Practitioner-observers make similar points. From one at National Semiconductor we hear, "People with degrees like a Harvard BA and a Stanford MBA last about seventeen months. They can't cope [with the flexibility and lack of structure]."

We recently encountered a very personal version of the business school complaint. When Dana's Rene McPherson, who had made his mark with sparkling accomplishment in that most difficult of areas, productivity in a slow-moving, unionized industry, became dean of the Stanford Graduate School of Business, one of our colleagues, who had just become associate dean, anxiously took us aside. "We've got to talk," he insisted. "I've just had my first long meeting with Ren. He talked to me about his Dana experience. Do you know that not one thing he did there is even mentioned in the MBA curriculum?"

THE MISSING PERSPECTIVE

The business schools, however, aren't running the country. Managers are. Underscoring the whole problem may be a missing perspective, the lack of any feeling for the whole on the part of the so-called professional manager. Again, Ed Wrapp makes the most forceful case:

The system is producing a horde of managers with demonstrable talents, but talents that are not in the mainstream of the enterprise. Professional managers are willing to study, analyze, and define the problem. They are steeped in specialization, standardization, efficiency, productivity, and quantification. They are highly rational and analytical. They insist on objective goals. . . . In some organizations, they can succeed if they are simply good at making presentations to the board of directors or writing strategies or plans. The tragedy is that these talents mask real deficiencies in overall management capabilities. These talented performers run for cover when grubby operating decisions must be made and often fail miserably when they are charged with earning a profit, getting things done and moving an organization forward.

Other observers have noted the same phenomenon. A *Business Week* writer, in a celebrated issue devoted to the subject of reindustrialization, put the case succinctly: most top management is "lacking a gut feeling for the gestalt of their businesses." Robert Hayes and William Abernathy, in a recent *Harvard Business Review* article, "Managing Our Way to Economic Decline," provide clues to the reason why: "No longer does the typical career . . . provide future top executives with intimate hands-on knowledge of the company's technologies, customers, and suppliers. . . . Since the mid-1950s there has been a rather substantial increase in the percentage of new company presidents whose primary interests and expertise lie in the financial and legal areas and not in production." And Hayes adds, "You don't have much of the spirit anymore of the top manager who simply looks at something and says, 'Damn it, this is a good product. Let's make it even though the payoff isn't apparent yet!'" Frederick Herzberg, another veteran observer of American management practice for more than forty years, declares simply: "Managers don't love the product. In fact, they are defensive about it."

By contrast, we have Japan's phenomenal success in cornering the small car market. What, exactly, is the nature of Japan's magic? *Fortune* suggests that it's not just gas mileage:

The Japanese deserve credit for far more than the circumstantial triumph of being able to supply efficient cars to a country [the United States] caught short of them. They excel in the quality of fits and finishes, moldings that match, doors that don't sag, materials that look good and wear well, flawless paint jobs. Most important of all, Japanese cars have earned a reputation for reliability, borne out by the generally lower rate of warranty claims they experience. Technically, most Japanese cars are fairly ordinary.

One of our favorite stories in support of *Fortune*'s analysis is about a Honda worker who, on his way home each evening, straightens up windshield wiper blades on all the Hondas he passes. He just can't stand to see a flaw in a Honda!

Now, why is all of this important? Because so much of excellence in performance has to do with people's being motivated by compelling, simple—even beautiful—values. As Robert Pirsig laments in *Zen and the Art of Motorcycle Maintenance:*

While at work I was thinking about this lack of care in the digital computer manuals I was editing. . . . They were full of errors, ambiguities, omissions and information so completely screwed up you had to read them six times to make any sense out of them. But what struck me for the first time was the agreement of these manuals with the spectator attitude I had seen in the shop. These were spectator manuals. It was built into the format of them. Implicit in every line is the idea that, "Here is the machine, isolated in time and space from everything else in the universe. It has no relationship to you, you have no relationship to it, other than to turn certain switches, maintain voltage levels, check for error conditions" and so on. That's it. The mechanics in their attitude toward the machine [Pirsig's motorcycle] were really taking no different attitude from the manual's toward the machine or from the attitude I had when I brought it in there. We were all spectators. It then occurred to me, there *is* no manual that deals with the *real* business of motorcycle maintenance, the most important aspect of all. Caring about what you're doing is considered either unimportant or taken for granted.

The attack next shifts to management's lack of concern for those people who might love the product if they were given the chance. To some critics, this charge sums it all up. Professor Abernathy recalls his surprise in discovering the reason for the Japanese success in autos: "The Japanese seem to have a tremendous cost advantage. . . . The big surprise to me was to find out that it's not automation. . . . They have developed a 'people' approach to the manufacturing of cars. . . . They have a work force that's turned on, willing to work, and is excited about making cars. . . . We have a different basic productivity position in this country, and it's because of a lot of minutiae. It's not the sort of thing that can be corrected by investment policy."

Steve Lohr takes up the cudgels on this point. He refers to chairman Akio Morita of Sony, who chides: "American managers are too little concerned about their workers." Morita goes on to describe his carefully designed revolution at Sony's U.S. plants. Lohr notes: "At Sony's plants in San Diego and Dothan, productivity has risen steadily, so that now it is very close to that of the company's factories in Japan." And Sony's highly publicized U.S. record pales next to Matsushita's post-purchase revival of Motorola's TV production operation. In five years, with virtually no replacement of the midwestern U.S. work force, the handful of Japanese general managers managed to cut the warranty bill from $22 million to $3.5 million, to cut defects per 100 sets from 140 to 6, to cut first ninety days (after sale) complaints from 70 percent to 7 percent, and to reduce personnel turnover from 30 percent a year to 1 percent.

Sony and Matsushita's success in the United States is a vivid reminder of the likely absence of any "Eastern magic" underpinning Japan's astounding productivity record. One commentator noted: "The productivity proposition is not so esoterically Japanese as it is simply human . . . loyalty, commitment through effective training, personal identification with the company's success and, most simply, the human relationship between the employee and his supervisor." There is one crucial cultural difference, however, that does seem to foster productivity through people in Japan. As a senior Japanese executive explained to us: "We are very different from the rest of the world. Our only natural resource is the hard work of our people."

Treating people—not money, machines, or minds—as the natural resource may be the key to it all. Kenichi Ohmae, head of McKinsey's Tokyo Office, says that in Japan *organization* and *people* (in the organization) are synonymous. Moreover, the people orientation encourages love of product and requires modest risk taking and innovation by the average worker. As Ohmae explains:

Japanese management keeps telling the workers that those at the frontier know the business best. . . . A well-run company relies heavily on individual or group initiatives for innovation and creative energy. The individual employee is utilized to the fullest extent of his creative and productive capacity. . . . The full organization—the proposal boxes, quality circles, and the like—looks "organic" and "entrepreneurial" as opposed to "mechanical" and "bureaucratic."

Kimsey Mann, the chief executive of Blue Bell, the world's second-largest apparel maker, referring to the eight attributes of management excellence on which this book is based, asserts that "every one of the eight is about people."

ANALYTIC IVORY TOWERS

The reason behind the absence of focus on product or people in so many American companies, it would seem, is the simple presence of a focus on something else. That something else is overreliance on analysis from corporate ivory towers and overreliance on financial sleight of hand, the tools that would appear to eliminate risk but also, unfortunately, eliminate action.

"A lot of companies overdo it," says Ed Wrapp. "They find planning more interesting than getting out a salable product. . . . Planning is a welcome respite from operating problems. It is intellectually more rewarding, and does not carry the pressures that operations entail. . . . Formal long-range planning almost always leads to overemphasis of technique." Fletcher Byrom of Koppers offers a suggestion. "As a regimen," he says, "as a discipline for a group of people, planning is very valuable. My position is, go ahead and plan, but once you've done your planning, put it on the shelf. Don't be bound by it. Don't use it as a major input to the decision-making process. Use it mainly to recognize change as it takes place." In a similar vein, *Business Week* recently reported: "Significantly, neither Johnson & Johnson, nor TRW, nor 3M—all regarded as forward thinking—has anyone on board called a corporate planner."

David Ogilvy, founder of Ogilvy and Mather, states bluntly: "The majority of businessmen are incapable of original thought because they are unable to escape from the tyranny of reason." Harvard's renowned marketing professor Theodore Levitt said recently: "Modelers build intricate decision trees whose pretension to utility is exceeded only by the awe in which high-level line managers hold the technocrats who construct them." Finally, we have a recent account of a Standard Brands' new product strategy that was an abject failure. The reason, according to a *Business Week* cover story, was that Standard Brands hired a bevy of GE planners and then gave them something akin to operating responsibility. After letting most of them go, the chairman noted: "The guys were bright, [but they] were not the kind of people who could implement the programs."

Now, all of this is apparently bad news for many who have made a life's work of number crunching. But the problem is not that companies ought not to plan. They damn well should plan. The problem is that the planning becomes an end in itself. It goes far beyond Byrom's sensible dictum to use it to enhance mental preparedness. Instead, the plan becomes the truth, and data that don't fit the preconceived plan (e.g., real customer response to a pre-test market action) are denigrated or blithely ignored. Gamesmanship replaces pragmatic action. ("Have you polled the corporate staffs yet about the estimate?" was a common query in one corporate operating committee that we observed for years.)

Business performance in the United States has deteriorated badly, at least compared to that of Japan, and sometimes to other countries—and in many cases absolutely, in terms of productivity and quality standards. We no longer make the best or most reliable products and we seldom make them for less, especially in internationally competitive industries (e.g., autos, chips).

The first wave of attack on the causes of this problem focused on government regulators. That, however, seemed to be an incomplete answer. Then, in mid-1980, the quest for root causes took thoughtful executives, business reporters, and academics alike into the heartland of management practice, all trying to figure out what had gone wrong. Not surprisingly, America's recent dependence on overanalysis and a narrow form of rationality bore the brunt of the attack. Both seemed especially at odds with the Japanese approach to the work force and to quality—even allowing for cultural differences.

The inquiry ran into two formidable roadblocks. The first was inherent defensiveness. The businessman's intellect and soul were finally under attack. Until then he had been encouraged by the press simply to increase his finger pointing at others, namely, the government. Second, the attack ran into a language problem. It wasn't seen as an attack on "a narrow form of rationality," what we have termed the "rational model," thereby calling for a broader form. It was seen as an attack on rationality and logical thought per se, thus implicitly encouraging escape into irrationality and mysticism. One was led to believe that the only solution was to move Ford board meetings to the local Zen center. And, obviously, that wasn't going to be the solution.

But let us stop for a moment and ask: What exactly do we mean by the fall of the rational model? We really are talking about what Thomas Kuhn, in his landmark book *The Structure of Scientific Revolutions*, calls a paradigm shift. Kuhn argues that scientists in any field and in any time possess a set of shared beliefs about the world, and for that time the set constitutes the dominant paradigm. What he terms "normal science" proceeds nicely under this set of shared beliefs. Experiments are carried out strictly within the boundaries of those beliefs and small steps toward progress are made. An old but excellent example is the Ptolemaic view of the universe (which held until the sixteenth century) that the earth was at the center of the universe, and the moon, sun, planets, and stars were embedded in concentric spheres around it. Elaborate mathematical formulas and models were developed that would accurately predict astronomical events based on the Ptolemaic paradigm. Not until Copernicus and Kepler found that the formula worked more easily when the sun replaced the earth as the center of it all did an instance of paradigm shift begin.

After a paradigm shift begins, progress is fast though fraught with tension. People get angry. New discoveries pour in to support the new belief system (e.g., those of Kepler and Galileo), and scientific revolution occurs. Other familiar examples of paradigm shift and ensuing revolution in science include the shift to relativity in physics, and to plate tectonics in geology. The important point in each instance is that the old "rationality" is eventually replaced with a new, different, and more useful one.

We are urging something of this kind in business. The old rationality is, in our opinion, a direct descendant of Frederick Taylor's school of scientific management and has ceased to be a useful discipline. Judging from the actions of managers who seem to operate under this paradigm, some of the shared beliefs include:

• Big is better because you can always get economies of scale. When in doubt, consolidate things; eliminate overlap, duplication, and waste. Incidentally, as you get big, make sure everything is carefully and formally coordinated.

* Low-cost producers are the only sure-fire winners. Customer utility functions lead them to focus on cost in the final analysis. Survivors always make it cheaper.
* Analyze everything. We've learned that we can avoid big dumb decisions through good market research, discounted cash-flow analysis, and good budgeting. If a little is good, then more must be better, so apply things like discounted cash flow to risky investments like research and development. Use budgeting as a model for long-range planning. Make forecasts. Set hard numerical targets on the basis of those forecasts. Produce fat planning volumes whose main content is numbers. (Incidentally, forget the fact that most long-range forecasts are bound to be wrong the day they are made. Forget that the course of invention is, by definition, unpredictable.)
* Get rid of the disturbers of the peace—i.e., fanatical champions. After all, we've got a plan. We want one new product development activity to produce the needed breakthrough, and we'll put 500 engineers on it if necessary, because we've got a better idea.
* The manager's job is decision making. Make the right calls. Make the tough calls. Balance the portfolio. Buy into the attractive industries. Implementation, or execution, is of secondary importance. Replace the whole management team if you have to to get implementation right.
* Control everything. A manager's job is to keep things tidy and under control. Specify the organization structure in great detail. Write long job descriptions. Develop complicated matrix organizations to ensure that every possible contingency is accounted for. Issue orders. Make black and white decisions. Treat people as factors of production.
* Get the incentives right and productivity will follow. If we give people big, straightforward monetary incentives to do right and work smart, the productivity problem will go away. Over-reward the top performers. Weed out the 30 to 40 percent dead wood who don't want to work.

- Inspect to control quality. Quality is like everything else; order it done. Triple the quality control department if necessary (forget that the QC force per unit of production in Japanese auto companies is just a third the size of ours). Have it report to the president. We'll show them (i.e., workers) that we mean business.
- A business is a business is a business. If you can read the financial statements, you can manage anything. The people, the products, and the services are simply those resources you have to align to get good financial results.
- Top executives are smarter than the market. Carefully manage the cosmetics of the income statement and balance sheet, and you will look good to outsiders. Above all, don't let quarterly earnings stop growing.
- It's all over if we stop growing. When we run out of opportunity in our industry, buy into industries we don't understand. At least we can then continue growing.

Much as the conventional business rationality seems to drive the engine of business today, it simply does not explain most of what makes the excellent companies work. Why not? What are its shortcomings?

For one, the numerative, analytical component has an in-built conservative bias. Cost reduction becomes priority number one and revenue enhancement takes a back seat. This leads to obsession with cost, not quality and value; to patching up old products rather than fooling with untidy new product or business development; and to fixing productivity through investment rather than revitalization of the work force. A buried weakness in the analytic approach to business decision making is that people analyze what can be most readily analyzed, spend more time on it, and more or less ignore the rest.

As Harvard's John Steinbruner observes, "If quantitative precision is demanded, it is gained, in the current state of things, only by so reducing the scope of what is analyzed that most of the important problems remain external to the analysis." This leads to fixation on the cost side of the equation. The numbers are "hardest" there. The fix, moreover, is mechanical and easy to picture—buy a new machine to replace nineteen jobs, reduce paperwork by 25 percent, close down two lines and speed up the remaining one.

Numerative analysis leads simultaneously to another unintended devaluation of the revenue side. Analysis has no way of valuing the extra oomph, the overkill, added by an IBM or Frito-Lay sales force. In fact, according to a recent observer, every time the analysts got their hands on Frito's "99.5 percent service level" (an "unreasonable" level of service in a so-called commodity business) their eyes began to gleam and they proceeded to show how much could be saved if only Frito would reduce its commitment to service. The analysts are "right"; Frito would immediately save money. But the analysts cannot possibly demonstrate the impact of a tiny degree of service unreliability on the heroic 10,000-person sales force—to say nothing of the Frito's retailers—and, therefore, on eventual market share loss or margin decline. Viewed analytically, the overcommitment to reliability by Caterpillar ("Forty-eight-hour parts service anywhere in the world—or Cat pays") or Maytag ("Ten years' trouble-free operation") makes no sense. Analytically, purposeful duplication of effort by IBM and 3M on product development, or cannibalization of one P&G brand by another P&G brand is, well, just that, duplication. Delta's family feeling, IBM's respect for the individual, and McDonald's and Disney's fetish for cleanliness make no sense in quantitative terms.

The exclusively analytic approach run wild leads to an abstract, heartless philosophy. Our obsession with body counts in Viet Nam and our failure to understand the persistence and long-time horizon of the Eastern mind culminated in America's most catastrophic misallocation of resources—human, moral, and material. But McNamara's fascination with numbers was just a sign of the times. One of his fellow whiz kids at Ford, Roy Ash, fell victim to the same affliction. Says *Fortune* of his Litton misadventures, "Utterly abstract in his view of business, [Ash] enjoyed to the hilt exercising his sharp mind in analyzing the most sophisticated accounting techniques. His brilliance led him to think in the most regal of ways: building new cities; creating a shipyard that would roll off the most

technically advanced vessels the way Detroit builds automobiles."
Sadly, *Fortune*'s analysis speaks not only of Ash's Litton failure but
also of the similar disaster ten years later that undid AM International under his leadership.

The rationalist approach takes the living element out of situations that should, above all, be alive. Lewis Lapham, the editor of *Harper's,* describes the fallacy of the numerative bias in an Easy Chair piece entitled "Gifts of the Magi": "The magi inevitably talk about number and weight—barrels of oil, the money supply—always about material and seldom about human resources; about things; not about people. The prevailing bias conforms to the national prejudice in favor of institutions rather than individuals." John Steinbeck made the same point about lifeless rationality:

The Mexican Sierra has 17 plus 15 plus 9 spines in the dorsal fin. These can easily be counted. But if the sierra strikes hard on the line so that our hands are burned, if the fish sounds and nearly escapes and finally comes in over the rail, his colors pulsing and his tail beating the air, a whole new relational externality has come into being—an entity which is more than the sum of the fish plus the fisherman. The only way to count the spines of the sierra unaffected by this second relational reality is to sit in a laboratory, open an evil-smelling jar, remove a stiff colorless fish from the formalin solution, count the spines and write the truth. . . . There you have recorded a reality which cannot be assailed—probably the least important reality concerning either the fish or yourself. . . . It is good to know what you are doing. The man with this pickled fish has set down one truth and recorded in his experience many lies. The fish is not that color, that texture, that dead, nor does he smell that way.

To be narrowly rational is often to be negative. Peter Drucker gives a good description of the baleful influence of management's analytic bias: "'Professional' management today sees itself often in the role of a judge who says 'yes' or 'no' to ideas as they come up. . . . A top management that believes its job is to sit in judgment will inevitably veto the new idea. It is always 'impractical.'" John Steinbruner makes a similar point commenting on the role of staffs in general: "It is inherently easier to develop a negative argument than to advance a constructive one." In his analysis of the MLF (NATO's proposed shared nuclear multi-lateral force) decision, Steinbruner recounts an exchange between a conservative academic and a real-world statesman. Secretary of State Dean Acheson said to the Harvard-trained presidential adviser Richard Neustadt, "You think Presidents should be warned. You're wrong. Presidents should be given confidence." Steinbruner goes on to analyze the roles of "warners" versus "bolsterers." Notwithstanding his attempt to present a balanced case, it is clear that the weight of the neutrally applied analytic model falls on the side of the warning, not the bolstering.

Mobil's chief executive, Rawleigh Warner, Jr., echoed the theme in explaining why his company decided not to bid on the 1960 offshore oil tracks in Prudhoe Bay: "The financial people in this company did a disservice to the exploration people. . . . The poor people in exploration were adversely impacted by people who knew nothing about oil and gas." Hayes and Abernathy, as usual, are eloquent on the subject: "We believe that during the past two decades American managers have increasingly relied on principles which prize analytical detachment and methodological elegance over insight . . . based on experience. Lacking hands-on experience, the analytic formulas of portfolio theory push managers even further toward an extreme of caution in allocating resources." Finally, George Gilder in *Wealth and Poverty* says, "Creative thought [the precursor to invention] requires an act of faith." He dissects example after example in support of his point, going back to the laying out of railroads, insisting that "when they were built they could hardly be justified in economic terms."

Today's version of rationality does not value experimentation and abhors mistakes. The conservatism that leads to inaction and years-long "study groups" frequently confronts businessmen with precisely what they were trying to avoid—having to make, eventually, one big bet. Giant product development groups analyze and analyze until years have gone by and they've designed themselves into one home-run product, with every bell and whistle attractive to every segment. Meanwhile, Digital, 3M, HP, and Wang, amid a hotbed of experimentation, have proceeded "irrationally" and chaotically, and introduced ten or more new products each during the same period. Advancement takes place only when we do something: try an early prototype on a customer or two, run a quick and dirty test market, stick a jury-rig device on an operating production line, test a new sales promotion on 50,000 subscribers.

The dominant culture in most big companies demands punishment for a mistake, no matter how useful, small, invisible. This is especially ironic because the most noble ancestor of today's business rationality was called *scientific* management. Experimentation is the fundamental tool of science: if we experiment successfully, by definition, we will make many mistakes. But overly rational businessmen are in pretty good company here, because even science doesn't own up to its messy road to progress. Robert Merton, a respected historian of science, describes the typical paper:

[There is a] rockbound difference between scientific work as it appears in print and the actual course of inquiry. . . . The difference is a little like that between textbooks of scientific method and the ways in which scientists actually think, feel, and go about their work. The books on methods present ideal patterns, but these tidy, normative patterns . . . do not reproduce the typically untidy, opportunistic adaptations that scientists really make. The scientific paper presents an immaculate appearance which reproduces little or nothing of the intuitive leaps, false starts, mistakes, loose ends, and happy accidents that actually cluttered up the inquiry.

Sir Peter Medawar, Nobel laureate in immunology, flatly declares, "It is no use looking to scientific 'papers,' for they do not merely conceal but actively misrepresent the reasoning which goes into the work they describe."

Anti-experimentation leads us inevitably to overcomplexity and inflexibility. The "home-run product" mentality is nowhere more evident than in the pursuit of the "superweapon" in defense. A *Village Voice* commentator notes:

The quickest way to understand the dread evoked in the Pentagon by Spinney [senior analyst with the Program Analysis and Evaluation division of the Department of the Defense] is to quote his bottom line: "Our strategy of pursuing ever-increasing technical complexity and sophistication has made high-technology solutions and combat readiness mutually exclusive." That is, the more money the U.S. presently spends on defense, the less able it is to fight. . . . More money has produced fewer but more complex planes which do not work much of the time. Deployment of fewer planes means a more elaborate and delicate communication system which is not likely to survive in war conditions.

Caution and paralysis-induced-by-analysis lead to an anti-experimentation bias. That, in turn, ironically leads to an ultimately risky "big bet" or the "superweapon" mentality. The screw turns once more. To produce such superproducts, hopelessly complicated and ultimately unworkable management structures are required. The tendency reaches its ultimate expression in the formal matrix organizational structure. Interestingly, some fifteen years before the mid-seventies matrix heyday, the researcher Chris Argyris identified the key matrix pathologies:

Why are these new administrative structures and strategies having trouble? . . . The assumption behind this [matrix] theory was that if objectives and critical paths to these objectives were defined clearly, people would tend to cooperate to achieve these objectives according to the best schedule they could devise. However, in practice, the theory was difficult to apply. . . . It was not long before the completion of the paperwork became an end in itself. Seventy-one percent of the middle managers reported that the maintenance of the product planning and program review paper flow became as crucial as accomplishing the line responsibility assigned to each group. . . . Another mode of adaptation was to withdraw and let the upper levels become responsible for the successful administration of the program. "This is their baby—let them make it work." . . . Still another frequently reported problem was the immobilization of the group with countless small decisions.

One can beat the complexity syndrome, but it is not easy. The IBM 360 is one of the grand product success stories in American business history, yet its development was sloppy. Along the way, chairman Thomas Watson, Sr., asked vice-president Frank Cary to "design a system to ensure us against a repeat of this kind of problem." Cary did what he was told. Years later, when he became chairman himself, one of his first acts was to get rid of the laborious product development structure that he had created for Watson. "Mr. Watson was right," he conceded. "It [the product development structure] will prevent a repeat of the 360 development turmoil. Unfortunately, it will also ensure that we don't ever invent another product like the 360."

The excellent company response to complexity is fluidity, the administrative version of experimentation. Reorganizations take place all the time. "If you've got a problem, put the resources on it and get it fixed," says one Digital executive. "It's that simple." Koppers's Fletcher Byrom adds support: "Of all the things that I have observed in corporations, the most disturbing has been a tendency toward over-organization, producing a rigidity that is intolerable in an era of rapidly accelerating change." HP's David Packard notes, "You've got to avoid having too rigid an organization. . . . If an organization is to work effectively, the communication should be through the most effective channel regardless of the organization chart. That is what happens a lot around here. I've often thought that after you get organized, you ought to throw the chart away." Speaking on the subject of American organizational rationality, our Japanese colleague Ken Ohmae says: "Most Japanese companies don't even have a reasonable organization chart. Nobody knows how Honda is organized, except that it uses lots of project teams and is quite flexible. . . . Innovation typically occurs at the interface, requiring multiple disciplines. Thus, the flexible Japanese organization has now, especially, become an asset."

The rationalist approach does not celebrate informality. Analyze, plan, tell, specify, and check up are the verbs of the rational process. Interact, test, try, fail, stay in touch, learn, shift direction, adapt, modify, and see are some of the verbs of the informal managing processes. We hear the latter much more often in our interviews with top performers. Intel puts in extra conference rooms, simply to increase the likelihood of informal problem solving among different disciplines. 3M sponsors clubs of all sorts specifically to enhance interaction. HP and Digital overspend on their own air and ground transportation systems just so people will visit one another. Product after product flows from Patrick Haggerty's bedrock principle of "tight coupling" at TI. It all means that people talk, solve problems, and fix things rather than posture, debate, and delay.

Unfortunately, however, management by edict feels more comfortable to most American managers. They shake their heads in disbelief at 3M, Digital, HP, Bloomingdale's, or even IBM, companies whose core processes seem out of control. After all, who in his right mind would establish Management By Wandering Around as a pillar of philosophy, as HP does? It turns out that the informal control through regular, casual communication is actually much tighter than rule by numbers, which can be avoided or evaded. But you'd have a hard time selling that idea outside the excellent companies.

The rational model causes us to denigrate the importance of values. We have observed few, if any, bold new company directions that have come from goal precision or rational analysis. While it is true that the good companies have superb analytic skills, we believe that their major decisions are shaped more by their values than by their dexterity with numbers. The top performers create a broad, uplifting, shared culture, a coherent framework within which charged-up people search for appropriate adaptations. Their ability to extract extraordinary contributions from very large numbers of people turns on the ability to create a sense of highly valued purpose. Such purpose invariably emanates from love of product, providing top-quality services, and honoring innovation and contribution from all. Such high purpose is inherently at odds with 30 quarterly MBO objectives, 25 measures of cost containment, 100 demeaning rules for production-line workers, or an ever-changing, analytically derived strategy that stresses costs this year, innovation next, and heaven knows what the year after.

There is little place in the rationalist world for internal competition. A company is not supposed to compete with itself. But throughout the excellent companies research, we saw example after example of that phenomenon. Moreover, we saw peer pressure—

rather than orders from the boss—as the main motivator. General Motors pioneered the idea of internal competition sixty years ago; 3M, P&G, IBM, HP, Bloomingdale's, and Tupperware are its masters today. Division overlap, product-line duplication, multiple new product development teams, and vast flows of information to spur productivity comparison—and improvements—are the watchwords. Why is it that so many have missed the message?

Again, the analyze–the–analyzable bias is ultimately fatal. It is true that costs of product-line duplication and nonuniformity of manufacturing procedures can be measured precisely. But the incremental revenue benefits from a steady flow of new products developed by zealous champions and the increment of productivity gains that comes from continuous innovation by competing shop floor teams are much harder, if not impossible, to get a handle on.

MISPLACED EMPHASIS

Perhaps the most important failing of the narrow view of rationality is not that it is wrong per se, but that it has led to a dramatic imbalance in the way we think about managing. Stanford's Harold Leavitt has a wonderful way of explaining this point. He views the managing process as an interactive flow of three variables: pathfinding, decision making, and implementation. The problem with the rational model is that it addresses only the middle element—decision making. In explaining the differences in the three activities, Leavitt has his classes first think of political leaders whose stereotypes most neatly fit the categories. For example, a typical class would suggest President John Kennedy as a pathfinder. For the decision-making stereotype, they might pick Robert McNamara in his role of Secretary of Defense or Jimmy Carter as President. For the prototypical implementer, everyone thinks of Lyndon Johnson ("Let us reason together," or "I'd rather have him inside the tent pissing out, than outside the tent pissing in.")

To add understanding, Leavitt has his class associate various occupations with his three categories. People who fall into the decision-making category include systems analysts, engineers, MBAs, statisticians, and professional managers—strange bedfellows, but very much alike in their bias for the rational approach. Implementing occupations would be those in which people essentially get their kicks from working with other people—psychologists, salesmen, teachers, social workers, and most Japanese managers. Finally, in the pathfinding category we find poets, artists, entrepreneurs, and leaders who have put their personal stamp on some business.

Obviously, the three processes are interconnected, and emphasis on any one trait to the exclusion of the other two is dangerous. The business ranks are full of would-be pathfinders—artists who can't get anything done. Likewise, implementers abound—compromising salesmen who have no vision. And the pitfalls of those who overemphasize decision making have been the subject of this chapter. The point of all this is that business management has at least as much to do with pathfinding and implementation as it does with decision making. The processes are inherently different, but they can complement and reinforce one another.

Pathfinding is essentially an aesthetic, intuitive process, a design process. There is an infinity of alternatives that can be posed for design problems, whether we are talking about architectural design or the guiding values for a business. From that infinity, there are plenty of bad ideas, and here the rational approach is helpful in sorting out the chaff. One is usually left with a large remaining set of good design ideas, however, and no amount of analysis will choose among them, for the final decision is essentially one of taste.

Implementation is also greatly idiosyncratic. As Leavitt points out, "People like their own children a lot, and typically aren't that interested in other people's babies." As consultants, we repeatedly find that it does the client no good for us to "analytically prove" that option A is the best—*and to stop at that point*. At that phase in the consulting process, option A is our baby, not theirs, and no amount of analytical brilliance is going to get otherwise uncommitted people to buy it. They have to get into the problem and understand it—and then own it for themselves.

As we've said, we don't argue for drastically tilting the balance toward either pathfinding or implementation. Rationality *is* important. A quality analysis will help to point a business in the right direction for pathfinding and will weed out the dumb options. But if America is to regain its competitive position in the world, or even hold what it has, we have to stop overdoing things on the rational side.

3

Man Waiting for Motivation

The central problem with the rationalist view of organizing people is that people are not very rational. To fit Taylor's old model, or today's organizational charts, man is simply designed wrong (or, of course, vice versa, according to our argument here). In fact, if our understanding of the current state of psychology is even close to correct, man is the ultimate study in conflict and paradox. It seems to us that to understand why the excellent companies are so effective in engendering both commitment and regular innovation from tens of thousands or even hundreds of thousands of people, we have to take into account the way they deal with the following contradictions that are built into human nature:

1. All of us are self-centered, suckers for a bit of praise, and generally like to think of ourselves as winners. But the fact of the matter is that our talents are distributed normally—none of us is really as good as he or she would like to think, but rubbing our noses daily in that reality doesn't do us a bit of good.

2. Our imaginative, symbolic right brain is at least as important as our rational, deductive left. We reason by stories *at least* as often as with good data. "Does it feel right?" counts for more than "Does it add up?" or "Can I prove it?"

3. As information processors, we are simultaneously flawed and wonderful. On the one hand, we can hold little explicitly in mind, at most a half dozen or so facts at one time. Hence there should be an

enormous pressure on managements—of complex organizations especially—to keep things very simple indeed. On the other hand, our unconscious mind is powerful, accumulating a vast storehouse of patterns, if we let it. Experience is an excellent teacher; yet most businessmen seem to undervalue it in the special sense we will describe.

4. We are creatures of our environment, very sensitive and responsive to external rewards and punishment. We are also strongly driven from within, self-motivated.

5. We act as if express beliefs are important, yet action speaks louder than words. One cannot, it turns out, fool any of the people any of the time. They watch for patterns in our most minute actions, and are wise enough to distrust words that in any way mismatch our deeds.

6. We desperately need meaning in our lives and will sacrifice a great deal to institutions that will provide meaning for us. We simultaneously need independence, to feel as though we are in charge of our destinies, and to have the ability to stick out.

Now, how do most companies deal with these conflicts? They take great pride in setting really high targets for people (productivity teams, product development teams, or division general managers), stretch targets. These are perfectly rational, but ultimately self-defeating. Why do TI and Tupperware, by contrast, insist that teams set their own objectives? Why does IBM set quotas so that almost all salespeople can make them? Surely TI has lazy workers. And no matter how intelligent IBM's hiring, screening, and training programs are for their salespeople, there is no way that this giant is going to get all superstars on its sales force. So what's going on?

The answer is surprisingly simple, albeit ignored by most managers. In a recent psychological study when a random sample of male adults were asked to rank themselves on "the ability to get along with others," *all* subjects, 100 percent, put themselves in the top half of the population. Sixty percent rated themselves in the top 10 percent of the population, and a full 25 percent ever so humbly thought they were in the top 1 percent of the population. In a parallel finding, 70 percent rated themselves in the top quartile in leadership; only 2 percent felt they were below average as leaders. Finally, in an area in which self-deception should be hard for most males, at least, 60 percent said they were in the top quartile of athletic ability; only 6 percent said they were below average.

We all think we're tops. We're exuberantly, wildly irrational about ourselves. And that has sweeping implications for organizing. Yet most organizations, we find, take a negative view of their people. They verbally berate participants for poor performance. (Most actually talk tougher than they act, but the tough talk nonetheless intimidates people.) They call for risk taking but punish even tiny failures. They want innovation but kill the spirit of the champion. With their rationalist hats on, they design systems that seem calculated to tear down their workers' self-image. They might not mean to be doing that, but they are.

The message that comes through so poignantly in the studies we reviewed is that we like to think of ourselves as winners. The lesson that the excellent companies have to teach is that there is no reason why we can't design systems that continually reinforce this notion; most of their people are made to feel that they are winners. Their populations are distributed around the normal curve, just like every other large population, but the difference is that their systems reinforce degrees of winning rather than degrees of losing. Their people by and large make their targets and quotas, because the targets and quotas are set (often by the people themselves) to allow that to happen.

In the not-so-excellent companies, the reverse is true. While IBM explicitly manages to ensure that 70 to 80 percent of its salespeople meet quotas, another company (an IBM competitor in part of its product line) works it so that only 40 percent of the sales force meets its quotas during a typical year. With this approach, at least 60 percent of the salespeople think of themselves as losers. They resent it and that leads to dysfunctional, unpredictable, frenetic behavior. Label a man a loser and he'll start acting like one. As one GM manager noted, "Our control systems are designed under the apparent assumption that 90 percent of the people are lazy ne'er-do-wells, just waiting to lie, cheat, steal, or otherwise screw us. We demoralize 95 percent of the work force who do act as adults by designing systems to cover our tails against the 5 percent who really are bad actors."

The systems in the excellent companies are not only designed to produce lots of winners; they are constructed to celebrate the winning once it occurs. Their systems make extraordinary use of non-monetary incentives. They are full of hoopla.

There are other opportunities for positive reinforcement. The most intriguing finding—in another major area of psychological research, called "attribution theory"—is the so-called fundamental attribution error postulated by Stanford's Lee Ross. Attribution theory attempts to explain the way we assign cause for success or failure. Was it good luck? Was it skill? Did we goof? Were we defeated by the system? The fundamental attribution error that so intrigues the psychologists is that we typically treat any success as our own and any failure as the system's. If anything goes well, it is quite clear that "I made it happen," "I am talented," and so on. If anything bad happens, "It's them," "It's the system." Once again, the implications for organizing are clear. People tune out if they feel they are failing, because "the system" is to blame. They tune in when the system leads them to believe they are successful. They learn that they can get things done because of skill, and, most important, they are likely to try again.

The old adage is "Nothing succeeds like success." It turns out to have a sound scientific basis. Researchers studying motivation find that the prime factor is simply the self-perception among motivated subjects that they are in fact doing well. Whether they are or not by any absolute standard doesn't seem to matter much. In one experiment, adults were given ten puzzles to solve. All ten were exactly the same for all subjects. They worked on them, turned them in, and were given the results at the end. Now, in fact, the results they were given were fictitious. Half of the exam takers were told that they had done well, seven out of ten correct. The other half were told they had done poorly, seven out of ten wrong. Then all were given another ten puzzles (the same for each person). The half who had been *told* that they had done well in the first round really did do better in the second, and the other half really did do worse. Mere association with past personal success apparently leads to more persistence, higher motivation, or something that makes us do better. Warren Bennis, in *The Unconscious Conspiracy: Why Leaders Can't Lead,* finds ample reason to agree: "In a study of school teachers, it turned out that when they held high expectations of their students, that alone was enough to cause an increase of 25 points in the students' IQ scores."

Research on the functions of the brain show that the left and right hemispheres differ substantially. The left half is the reasoning, sequential, verbal half; it is the "logical" and rational half. The right half is the artistic half; it is the half that sees and remembers patterns, recalls melodies, waxes poetic. The utter distinctness of the two hemispheres has been shown repeatedly, when, for example, required surgery in cases of *grand mal* epilepsy has decoupled the links between the two halves. Studies show that the right half is great at visualizing things but can't verbalize any of them. The left side can't remember patterns, like people's faces. Those who say "I'm no good at names, but never forget a face" aren't defective; simply a little right-brained.

Arthur Koestler points out the dominant role, like it or not, of our right brain. In his *Ghost in the Machine* Koestler attributes our basest emotions, our predilection for war and destruction, to "an underdeveloped [right] half of the brain." He asserts that "[our] behavior continues to be dominated by a relatively crude and primitive system." And Ernest Becker goes so far as to say that "the psychoanalytic emphasis on creatureliness [i.e., our basic traits] is *the* lasting insight on human character." He adds that it leads us urgently to "seek transcendence," "avoid isolation," and "above all fear helplessness."

The organizational implications of this line of reasoning are inescapable, although with a potential dark side (e.g., we'll do almost anything to seek transcendence). The business researcher Henry Mintzberg amplifies the point:

One fact recurs repeatedly in all of this research: the key managerial processes are enormously complex and mysterious (to me as a researcher, as well as to the managers who carry them out), drawing on the vaguest of information and using the least articulated of mental processes. These processes seem to be more relational and holistic than ordered and sequential, and more intuitive than intellectual; they seem to be most characteristic of right-hemispheric activity.

The total of left- and right-brain research suggests simply that businesses are full (100 percent) of highly "irrational" (by left-brain standards), emotional human beings: people who want desperately to be on winning teams ("seek transcendence"); individuals who thrive on the camaraderie of an effective small group or unit setting ("avoid isolation"); creatures who want to be made to feel that they are in at least partial control of their destinies ("fear helplessness"). Now, we seriously doubt that the excellent companies have explicitly proceeded from right-brain considerations in developing their management practices. But the effect is such that it appears they have, especially in relation to their competitors. They simply allow for—and take advantage of—the emotional, more primitive side (good and bad) of human nature. They provide an opportunity to be the best, a context for the pursuit of quality and excellence. They offer support—more, celebration; they use small, intimate units (from divisions to "skunk works" or other uses of teams); and they provide within protected settings opportunities to stand out—as part of a quality circle at TI, for example, where there are 9,000 such entities.

Also note that this implicit recognition of the right-side traits by the excellent companies is directly at the expense of more traditional left-brain business practices: causes to fight for are a long way from thirty quarterly MBO objectives. The intimate team or small division ignores scale economies. Allowing freedom of expression by thousands of quality circles flies in the face of the "one best way" of traditional production organization.

There is another aspect to our right brain's nature that isn't usually a part of conventional management wisdom but is clearly being nurtured by the excellent companies. That is the intuitive, creative side. Science and mathematics are thought by many to be the mecca of logical thought, and logical, rational thought certainly does feature prominently in the day-to-day progression of science. But as we pointed out in connection with scientific paradigm change, logic is not the true engine of scientific progress. Here's how James Watson, co-discoverer of the structure of DNA, described the double helix the night he finished his research: "It's so beautiful, you see, so beautiful." In science the aesthetic, the beauty of the concept, is so important that Nobel laureate Murray Gell-Mann was moved to comment, "When you have something simple that agrees with all the rest of the physics and really seems to explain what's going on, a few experimental data against it are no objection whatsoever." When McDonald's former chairman Ray Kroc waxed poetic about hamburger buns, he hadn't taken leave of his senses; he simply recognized the importance of beauty as a starting point for the business logic that ensues.

We "reason" with our intuitive side just as much as, and perhaps more than, with our logical side. Two experimental psychologists, Amos Tversky and Daniel Kahneman, are the leaders of a principal thrust of experimental psychology called "cognitive biases," started about fifteen years ago. In test after test, with sophisticated—even scientifically trained—subjects, our bias for the intuitive manifests itself. For example, a phenomenon they term "representativeness" strongly affects our reasoning powers. Simply said, we are more influenced by stories (vignettes that are whole and make sense in themselves) than by data (which are, by definition, utterly abstract). In a typical experiment, subjects are told a story about an individual, given some relevant data, and then asked to guess the individual's career. The subjects are told, say, "Jack is a forty-five-year-old man. He is married and has four children. He is generally conservative, careful, and ambitious. He shows no interest in political and social issues and spends most of his free time on his many hobbies, which include home carpentry, sailing, and mathematical puzzles." Then the subjects are told that Jack's description was selected from a population that contains 80 percent lawyers and 20 percent engineers. It doesn't matter that they are told the sample is lawyer-heavy; the subjects pick the occupation on the basis of their stereotype of the occupation. In this case, most of the subjects decided that Jack was an engineer.

Gregory Bateson also states the case for the primacy of representativeness:

There's a story which I have used before and shall use again: a man wanted to know about mind, not in nature, but in his private large computer. He asked it, "Do you compute that you will ever think like a human being?" The machine then set to work to analyze its own computational habits. Finally, the machine printed its answer on a piece of paper, as such machines do. The man ran to get the answer and found, neatly typed, the words: THAT REMINDS ME OF A STORY. A story is a little knot or complex of that species of connectedness which we call relevance. Surely the computer was right. This is indeed how people think.

Related findings include:

1. We don't pay attention to prior outcomes. History doesn't move us as much as does a good current anecdote (or, presumably, a juicy bit of gossip). We reason with data that come readily to mind (called the "availability heuristic" by Kahneman and Tversky) even if the data have no statistical validity. When we meet three friends in the space of a week in a hotel in Tokyo, we are more apt to think "how odd" than we are to muse on the probability that our circle of acquaintances tends to frequent the same places we do.

2. If two events even vaguely co-exist, we leap to conclusions about causality. For example, in one experiment subjects are given clinical data on people and drawings of them. Later, when asked to recall what they have found, they will greatly overestimate the correlation between the way a person looks and that person's true characteristics—people who are in fact suspicious by nature were judged typically (and erroneously) to have peculiar eyes.

3. We're hopeless about sample size. We find small samples

about as convincing as large ones, sometimes more so. Consider, for example, a situation in which an individual draws two balls from an urn and finds that both are red. Another person then draws thirty balls and finds that eighteen are red and twelve are white. Most people believe the first sample contains the stronger evidence that the urn contains predominantly red balls, although as a purely statistical matter, the opposite is the case.

And so it goes through a wealth of experimental data, now thousands of experiments old, showing that people reason intuitively. They reason with simple decision rules, which is a fancy way of saying that, in this complex world, they trust their gut. We need ways of sorting through the infinite minutiae out there, and we start with heuristics—associations, analogues, metaphors, and ways that have worked for us before.

There is both good and bad in this, although mainly good, we think. The bad part is that, as the experiments demonstrate, our collective gut is not much use in the arcane world of probability and statistics. Here is an area in which a little more training on the rational side would help! But the good element is that it probably is only the intuitive leap that will let us solve problems in this complex world. This is a major advantage of man over computer, as we will see.

SIMPLICITY AND COMPLEXITY

Most acronyms stink. Not KISS: Keep It Simple, Stupid! One of the key attributes of the excellent companies is that they have realized the importance of keeping things simple despite overwhelming genuine pressures to complicate things. There is a powerful reason for this, and we turn to the Nobel laureate Herbert Simon for the answer. Simon has been deeply involved in the field of artificial intelligence in recent years, trying to get computers to "think" more as people do rather than conducting inefficient, exhaustive searches for solutions.

Among Simon and his colleagues' most important findings, for example, is that human beings are not good at processing large streams of new data and information. They have found that the most we can hold in short-term memory, without forgetting something, is six or seven pieces of data.

Again, we're faced with an important paradox for management, for the world of big companies *is* complex. Just how complex is suggested by the fact that as the number of people in a company goes up arithmetically, the number of possible interactions among them goes up geometrically. If our company has ten employees, we can all stay in touch with one another because the number of ways we can interact, say, in one-on-one discussions is forty-five. If our company has 1,000 employees, on the other hand, that same number of possible one-on-one interactions goes up to about 500,000. If there are 10,000 employees then the number rises to 50 million. To cope with the complex communications needs generated by size alone, we require appropriately complex systems, or so it would seem.

We recently read a stack of business proposals, none of which was less than fifty pages long. We subsequently went through the personal programs of the senior executives of a $500 million consumer goods company; seldom did its programs contain fewer than fifteen objectives for the year, and thirty objectives was not uncommon. Not unreasonable, you say, until you realize that the team at the top is trying to keep informed on the career progress of the top 500 people in the company—perhaps 15,000 objectives. Now, what's the logical response to things getting more and more complex for the top executives? What do they do when they start getting thousands of objectives they somehow are expected to process? What do they do when all these objectives are but a tiny part of the total set of information they must deal with? Well, they hire staff to simplify things for themselves.

Staff may, in fact, simplify matters—for them. But the staff makes life miserable for the people in the field. The moment that staff, in any number, leaps into action, it starts generating information requests, instructions, regulations, policies, reports, and finally questionnaires on "how staff is doing." Somewhere along the way to bigness, information overload sets in. Short-term memory can't process it all, or even a small fraction of it, and things get very confusing.

But, as is so often the case, the excellent companies seem to have found ways of coping with this problem. For one thing, they intentionally keep corporate staffs small. Then there aren't enough corporate staff around to generate too much confusion down the line. Emerson, Schlumberger, and Dana, for example, are $3 billion to $6 billion top-performing corporations; yet each is run with fewer than 100 bodies in corporate headquarters. Ford, meanwhile, has seventeen layers of management, while Toyota (and the 800 million-member Roman Catholic Church) has five. As another coping device, the excellent companies focus on only a few key business values, and a few objectives. The focus on a few key values lets everyone know what's important, so there is simply less need for daily instructions (i.e., daily short-term memory overload). Rene McPherson, when he took over at Dana, dramatically threw out 22½ inches of policy manuals and replaced them with a one-page statement of philosophy focusing on the "productive people." (His auditors were appalled. "That means there could be seventy-four different procedures in seventy-four different plants." McPherson replied, "Yes, and it means maybe you guys will finally have to earn your fees.")

Many of these companies eliminate paperwork through their use of quick-hit task forces, and among the paperwork fighters P&G is legendary for its insistence on one-page memos as the almost sole means of written communication. Others "suboptimize"; they ignore apparent economies of scale, putting up with a fair amount of internal overlap, duplication, and mistakes just so they won't have to coordinate everything, which, given their size, they couldn't do anyway. As we go through the research results in later chapters, we shall find scores of devices used by the excellent companies for keeping things simple. In every instance they are ignoring the "real world," the complex one. They are, in a real sense, being *simplistic*, not just keeping it simple. Of course, "More than two objectives is no objectives," the TI watchword, is unrealistic; thirty objectives *is* a more realistic description of the world. But the TI rule jibes with human nature. With a little luck and a hell of a lot of persistence, one might actually get two things done in a year.

Simon, in his research on artificial intelligence, finds another fascinating result that is, finally, encouraging. Looking at long-term memory, he and his colleagues studied the problem of programming computers to play chess. Within this research lies an important idea that ties together the role of the rational and the role of the intuitive. Simon started by assuming that the game of chess could be played on a strictly rationalist basis, that is, one could program the computer like a decision tree. Before moving, the computer would search ahead and examine all possible moves and countermoves. Theoretically, that can be done. However, it's not practical, for the number of possibilities is something on the order of 10 to the 120th power (a trillion, by contrast, is only 10 to the 12th power). The fastest of modern-day computers can do something like 10 to the 20th calculations in a century. So programming our chess-playing computer to behave rationally is just not feasible.

Struck by the notion, Simon went on to research what good chess players really do. In conducting his research, he asked chess masters—the best in the world—to look briefly (for ten seconds) at games that were already in progress, the boards still containing around twenty or so pieces. He found that the chess masters could later recall the locations of virtually all the pieces. That doesn't fit with short-term memory theory at all. When class A players (one rank below masters) were asked to do the same test, they scored much less well. Maybe chess masters have better short-term memories. But here's the rub with that idea: neither the masters nor the class A players could remember where the pieces were on chessboard set-ups that were randomly generated without games in progress. Something else must be at work.

The something else, Simon believes, is that the chess masters have much more highly developed long-term chess memories, and the memories take the form of subconsciously remembered patterns, or what Simon terms chess "vocabularies." While the class A player has a vocabulary of around 2,000 patterns, the chess master

has a vocabulary of around 50,000 patterns. Chess players use deci-
sion-tree thinking, it appears, only in a very limited sense. They
begin with the patterns: Have I seen this one before? In what context? What worked before?

When we start to dwell on the implications of Simon's research,
we are struck by its applicability elsewhere. The mark of the true
professional in any field is the rich vocabulary of patterns, developed through years of formal education and especially through
years of practical experience. The experienced doctor, the artist, the
machinist, all have rich pattern vocabularies—Simon is now calling
them "old friends."

This notion ought to be celebrated for, in our minds, it is the real
value of experience in business. It helps to explain the importance
of management by wandering around. Not only do the employees
benefit from being paid attention to. The experienced boss has good
instincts; his vocabulary of old-friend patterns tells him immediately whether things are going well or badly.

The vocabulary-of-patterns notion ought to do several things for
us as we think about its implications for excellence in management.
It should help us trust our gut more often on key business decisions.
It should lead us to ask the advice of customers and workers more
frequently. And finally it should encourage all of us to think hard
about the value of experimenting as opposed to merely detached
study.

POSITIVE REINFORCEMENT

B. F. Skinner has a bad reputation in some circles. His techniques
are seen as ultimately manipulative. He actually sets himself up for
attack from all quarters. In his popular treatise *Beyond Freedom
and Dignity,* for instance, he calls for nothing less than a sweeping
"technology of behavior." He says that we are all simply a product
of the stimuli we get from the external world. Specify the environment completely enough and you can exactly predict the individual's actions. We are confronted with the same problem that the
rationalists ran into with economic man. Just as economic man can

never know enough (i.e., everything) to maximize his utility function, we can't ever come close to specifying the environment completely enough to predict behavior. Unfortunately, though, we tend to throw out some of Skinner's extremely powerful and practical findings because of the arrogance of his claims and the implicit ideology associated with them

If we look further, we find that the most important lesson from Skinner is the role of positive reinforcement, of rewards for jobs well done. Skinner and others take special note of the asymmetry between positive and negative reinforcement (essentially the threat of sanctions). In short, negative reinforcement will produce behavioral change, but often in strange, unpredictable, and undesirable ways. Positive reinforcement causes behavioral change too, but usually in the intended direction.

Why spend time on this? It seems to us that central to the whole notion of managing is the superior/subordinate relationship, the idea of manager as "boss," and the corollary that orders will be issued and followed. The threat of punishment is the principal implied power that underlies it all. To the extent that this underlying notion prevails, we are not paying attention to people's dominant need to be winners. Moreover, repeated negative reinforcement is, as Skinner says, usually a dumb tactic. It doesn't work very well. It usually results in frenetic, unguided activity. Further, punishment doesn't suppress the desire to "do bad." Says Skinner: "The person who has been punished is not thereby simply less inclined to behave in a given way; at best, he learns how to avoid punishment."

Positive reinforcement, on the other hand, not only shapes behavior but also teaches and in the process enhances our own self-image. To give a negative example first, suppose that we get punished for "not treating a customer well." Not only do we not know what specifically to do in order to improve; we might well respond by "learning" to avoid customers altogether. In Skinner's terms, "customer" per se, rather than "treating a customer badly," has become associated with punishment. On the other hand, if someone tells us via a compliment from a "mystery shopper" that we "just acted in the best traditions of XYZ Corporation in responding to Mrs.

Jones's minor complaint," well, that's quite different. Per Skinner, and our own experience, what we are now likely to get is an employee out beating the bushes to find more Mrs. Joneses to treat well. He or she has learned that a specific (positive) behavior pattern leads to rewards and has at the same time satisfied the insatiable human need to enhance one's self-image.

Heinz's highly successful frozen foods subsidiary, Ore-Ida, is trying an intriguing variation on this theme in order to encourage more learning and risk taking in its research activities. It has carefully defined what it calls the "perfect failure," and has arranged to shoot off a cannon in celebration every time one occurs. The perfect failure concept arises from simple recognition that all research and development is inherently risky, that the only way to succeed at all is through lots of tries, that management's primary objective should be to induce lots of tries, and that a good try that results in some learning is to be celebrated even when it fails. As a by-product, they legitimize and even create positive feelings around calling a quick halt to an obviously failing proposition, rather than letting it drag on with resulting higher cost in funds and eventual demoralization.

Positive reinforcement also has an intriguing Zen-like property. *It nudges good things onto the agenda instead of ripping things off the agenda.* Life in business, as otherwise, is fundamentally a matter of attention—how we spend our time. Thus management's most significant output is getting others to shift attention in desirable directions (e.g., "Spend more time in the field with customers"). There are only two ways to accomplish such a shift. First, we attempt through positive reinforcement to lead people gently over a period of time to pay attention to new activities. This is a subtle shaping process. Or we can "take the bull by the horns" and simply try to wrestle undesirable traits off the agenda (e.g., "Quit staying in the office filling in forms"). Skinner's argument is that the wrestler's approach is likely to be much less efficient, even though it may not seem that way in the very short run. That is, ripping items off an agenda leads to either overt or covert resistance: "I'll get out of the office, if you insist, but I'll spend the time in the local pub."

The "nudge it on the agenda" approach leads to a natural diffusion process. The positively reinforced behavior slowly comes to occupy a larger and larger share of time and attention. By definition, *something* (who cares what?) less desirable begins to drop off the agenda. But it drops off the agenda on the basis of our sorting process. The stuff that falls off is what we want to push off in order to make room for the positively reinforced items. The difference in approach is substantial. If, by force of time alone (a nonaversive force), *we* choose to push a low-priority item off, then it is highly unlikely that we will cheat on ourselves and try to do more of the less attractive (just pushed off the agenda) behavior. So, back to Zen; the use of positive reinforcement goes with the flow rather than against it.

Our general observation is that most managers know very little about the value of positive reinforcement. Many either appear not to value it at all, or consider it beneath them, undignified, or not very macho. The evidence from the excellent companies strongly suggests that managers who feel this way are doing themselves a great disservice. The excellent companies seem not only to know the value of positive reinforcement but how to manage it as well.

As Skinner notes, the way the reinforcement is carried out is more important than the amount. First, it ought be *specific*, incorporating as much information content as possible. We note, for instance, that activity-based MBO systems ("Get the Rockville plant on line by July 17") are more common in the excellent companies than are financially based MBOs.

Second, the reinforcement should have *immediacy*. Thomas Watson, Sr., is said to have made a practice of writing out a check on the spot for achievements he observed in his own peripatetic management role. Other examples of on-the-spot bonuses were mentioned frequently in our research. At Foxboro, a technical advance was desperately needed for survival in the company's early days. Late one evening, a scientist rushed into the president's office with a working prototype. Dumbfounded at the elegance of the solution and bemused about how to reward it, the president bent forward in his chair, rummaged through most of the drawers in his desk, found something, leaned over the desk to the scientist, and said, "Here!"

In his hand was a banana, the only reward he could immediately put his hands on. From that point on, the small "gold banana" pin has been the highest accolade for scientific achievement at Foxboro. Lest that seem too mundane, at HP we unearthed a tale of marketers anonymously sending pound bags of pistachios to a salesman who sold a new machine.

Third, the system of feedback mechanisms should take account of *achievability*. Major gold banana events are not common, so the system should reward small wins. Good news swapping is common in the excellent companies.

The fourth characteristic is that a fair amount of the feedback comes in the form of *intangible* but ever-so-meaningful attention from top management. When you think about it, with management's time being as scarce as it is, that form of reinforcement may be the most powerful of all.

Finally, Skinner asserts that regular reinforcement loses impact because it comes to be expected. Thus *unpredictable* and *intermittent* reinforcements work better—the power of walking the shop floor again. Moreover, small rewards are frequently more effective than large ones. Big bonuses often become political, and they discourage legions of workers who don't get them but think they deserve them. Remember, we all think we're winners. Have you ever been around a member of a product launch team who didn't think that it was really his personal contribution that turned the tide in getting the new widget out the door? The small reward, the symbolic one, becomes a cause for positive celebration rather than the focus of a negative political battle.

Skinner's reinforcement notions have many offshoots. Arguably the most important is Leon Festinger's now widely held "social comparison theory." His hypothesis, presented in 1951, was simply that people most strenuously seek to evaluate their performance by comparing themselves to others, not by using absolute standards. (Actually, this line of inquiry has origins going back to 1897, when Norman Triplett observed in a controlled experiment that bicyclists "race faster against each other than against a clock.") We see many evidences of the use of social comparison by the excellent companies. Among them are regular peer reviews (the mainstay of the TI, Intel, and Dana management systems); information made widely available on comparative performance (sales groups, tiny productivity teams, and the like); and purposefully induced internal competition (for example, among P&G brand managers). All are practices that stand in marked contrast to the conventional management techniques. As a young man, Rene McPherson was nearly fired in 1955 for telling people in his plant what their sales and profits were and how they stacked up against other plants. In 1972, as Dana's chairman, he visited a Toledo plant, open since 1929, where managers and employees had never been exposed to performance information. This tale is sadly not exceptional. We expect people to be motivated in a vacuum.

To put things into proper perspective, however, we should stress that we are *not* advocating reinforcement as the starting point for theory on what makes excellent companies tick. Skinner's work is important, and, as we said, underutilized in most management theory and practice. But the larger context of high performance, we believe, is *intrinsic motivation*. On the surface, self-motivation is opposed in many ways to the beliefs of reinforcement theory; but in our minds the two contexts fit together nicely. In experiment after experiment, Edward Deci of the University of Rochester has shown that lasting commitment to a task is engendered only by fostering conditions that build intrinsic motivations. In plain talk, Deci finds that people must believe that a task is inherently worthwhile if they really are to be committed to it. (In addition, he also finds that if we too regularly reward a task, we often vitiate commitment to it.)

It may not be surprising that managers have not taken a shine to the use of positive reinforcement. It smacks of the Brave New World, on the one hand (too tough), and of arbitrary back patting on the other (too soft). However, we are surprised at the degree to which intrinsic motivation has been underutilized in most companies. The excellent companies, by contrast, tap the inherent worth of the task as a source of intrinsic motivation for their employees. TI and Dana insist that teams and divisions set their own goals. Virtually all of the excellent companies are driven by just a few key values, and then give lots of space to employees to take initiatives in support of those values—finding their own paths, and so making the task and its outcome their own.

ACTION, MEANING, AND SELF-CONTROL

Probably few of us would disagree that actions speak louder than words, but we behave as if we don't believe it. We behave as if the proclamation of policy and its execution were synonymous. "But I made quality our number one goal years ago," goes the lament. Managers can't drive forklifts any more. Yet they do still act. They do *something*. In short, they pay attention to some things and not to others. Their action expresses their priorities, and it speaks much louder than words. In the quality case alluded to above, a president's subordinate clarified the message, "Of course, he's for quality. That is, he's never said, 'I don't care about quality.' It's just that he's for everything. He says, 'I'm for quality,' twice a year and he acts, 'I'm for shipping product,' twice a day." In another case, a high technology company president pinned his company's revitalization hopes on new products, publicly proclaiming (e.g., to the securities analysts) that they were on the way. A look at his calendar and phone log revealed that only 3 percent of his time was actually spent on new products. Yet he kept asking us in all sincerity why even his closest allies weren't getting the message.

Intriguingly, this ambiguous area is a subject of heated long-term debate in psychology. There are two schools of thought. One says that attitudes (beliefs, policies, proclamations) precede actions— the "Tell, then do" model. The other, clearly more dominant, reverses the logic. The Harvard psychologist Jerome Bruner captures the spirit when he says, "You more likely act yourself into feeling than feel yourself into action." A landmark experiment, carried out in 1934, spurred the controversy. It demonstrated unequivocally that there is often little relationship between explicitly stated belief and mundane action:

LaPiere, a white professor, toured the United States in 1934 with a young Chinese student and his wife. They stopped at 66 hotels or motels and at 184 restaurants. All but one of the hotels or motels gave them space, and they were never refused service at a restaurant. Sometime later a letter was sent to these establishments asking whether they would accept Chinese as guests. [There was a strong anti-Chinese bias in the United States at the time.] Ninety-two percent said they would not. LaPiere, and many after him, interpreted these findings as reflecting a major inconsistency between behavior and attitudes. Almost all the proprietors *behaved* in a tolerant fashion, but they expressed an intolerant *attitude* when questioned by letter.

Analogously, what's called "foot-in-the-door research" demonstrates the importance of incrementally acting our way into major commitment. For instance, in one experiment, in Palo Alto, California, most subjects who initially agreed to put a *tiny* sign in their front window supporting a cause (traffic safety) subsequently agreed to display a billboard in their front yard, which required letting outsiders dig sizable holes in the lawn. On the other hand, those not asked to take the first small step turned down the larger one in ninety-five cases out of a hundred.

The implications of this line of reasoning are clear: only if you get people *acting,* even in small ways, the way you want them to, will they come to believe in what they're doing. Moreover, the process of enlistment is enhanced by explicit *management* of the after-the-act labeling process—in other words, publicly and ceaselessly lauding the small wins along the way. "Doing things" (lots of experiments, tries) leads to rapid and effective learning, adaptation, diffusion, and commitment; it is the hallmark of the well-run company.

Moreover, our excellent companies appear to do their way into strategies, not vice versa. A leading researcher of the strategic process, James Brian Quinn, talks about the role of leadership in strategy building. It doesn't sound much like a by-the-numbers, analysis-first process. He lists major leadership tasks, and the litany includes amplifying understanding, building awareness, changing symbols, legitimizing new viewpoints, making tactical shifts and testing partial solutions, broadening political support, overcoming opposition, inducing and structuring flexibility, launching trial balloons and engaging in systematic waiting, creating pockets of commitment, crystallizing focus, managing coalitions, and formalizing commitment (e.g., empowering "champions"). The role of the leader, then, is one of orchestrator and labeler: taking what can be gotten in the way of action and shaping it—generally after the fact—into lasting commitment to a new strategic direction. In short, he makes meanings.

The leading mathematician Roger Penrose says, "The world is an illusion created by a conspiracy of our senses." Yet we poor mortals try valiantly, at times desperately, to inscribe meaning on the *tabula rasa* given to us at birth. As Bruno Bettelheim has observed in *On the Uses of Enchantment,* "If we hope to live not just from moment to moment, but in true consciousness of our existence, then our greatest need and most difficult achievement is to find meaning in our lives." Bettelheim emphasizes the historically powerful role of fairy tales and myths in shaping meaning in our lives.

As we worked on research of our excellent companies, we were struck by the dominant use of story, slogan, and legend as people tried to explain the characteristics of their own great institutions. All the companies we interviewed, from Boeing to McDonald's, were quite simply rich tapestries of anecdote, myth, and fairy tale. And we do mean fairy tale. The vast majority of people who tell stories today about T. J. Watson of IBM have never met the man or had direct experience of the original more mundane reality. Two HP engineers in their mid-twenties recently regaled us with an hour's worth of "Bill and Dave" (Hewlett and Packard) stories. We were subsequently astonished to find that neither had seen, let alone talked to, the founders. These days, people like Watson and A. P. Giannini at Bank of America take on roles of mythic proportions that the real persons would have been hard-pressed to fill. Nevertheless, in an organizational sense, these stories, myths, and legends appear to be very important, because they convey the organization's shared values, or culture.

Without exception, the dominance and coherence of culture proved to be an essential quality of the excellent companies. Moreover, the stronger the culture and the more it was directed toward the marketplace, the less need was there for policy manuals, organization charts, or detailed procedures and rules. In these companies, people way down the line know what they are supposed to do in most situations because the handful of guiding values is crystal clear. One of our colleagues is working with a big company recently thrown together out of a series of mergers. He says: "You know, the problem is *every* decision is being made for the first time. The top people are inundated with trivia because there are no cultural norms."

By contrast, the shared values in the excellent companies are clear, in large measure, because the mythology is rich. Everyone at Hewlett-Packard knows that he or she is supposed to be innovative. Everyone at Procter & Gamble knows that product quality is the sine qua non. In his book on P&G, *Eyes on Tomorrow,* Oscar Schisgall observes: "They speak of things that have very little to do with price of product. . . . They speak of business integrity, of fair treatment of employees. 'Right from the start,' said the late Richard R. Deupree when he was chief executive officer, 'William Procter and James Gamble realized that the interests of the organization and its employees were inseparable. That has never been forgotten.'"

Poorer-performing companies often have strong cultures, too, but dysfunctional ones. They are usually focused on internal politics rather than on the customer, or they focus on "the numbers" rather than on the product and the people who make and sell it. The top companies, on the other hand, always seem to recognize what the companies that set only financial targets don't know or don't deem important. The excellent companies seem to understand that *every* man seeks meaning (not just the top fifty who are "in the bonus pool").

Perhaps transcendence is too grand a term for the business world, but the love of product at Cat, Bechtel, and J&J comes very close to meriting it. Whatever the case, we find it compelling that so many thinkers from so many fields agree on the dominating need of human beings to find meaning and transcend mundane things. Nietzsche believed that "he who has a *why* to live for can bear almost any *how*." John Gardner observes in *Morale,* "Man is a stubborn seeker of meaning."

Some of the riskiest work we do is concerned with altering organization structures. Emotions run wild and almost everyone feels threatened. Why should that be? The answer is that if companies do not have strong notions of themselves, as reflected in their values, stories, myths, and legends, people's only security comes from where they live on the organization chart. Threaten that, and in the absence of some grander corporate purpose, you have threatened the closest thing they have to meaning in their business lives.*

So strong is the need for meaning, in fact, that most people will yield a fair degree of latitude or freedom to institutions that give it to them. The excellent companies are marked by very strong cultures, so strong that you either buy into their norms or get out. There's no halfway house for most people in the excellent companies. One very able consumer marketing executive told us, "You know, I deeply admire Procter & Gamble. They are the best in the business. But I don't think I could ever work there." She was making the same point that Adam Myerson at *The Wall Street Journal* had in mind when he urged us to write an editorial around the theme: "Why we wouldn't want to work for one of our excellent companies." The cultures that make meanings for so many repel others.

Some who have commented on our research wonder if there is not a trap or two in the very strength of the structures and cultures of the well-run companies. There probably is. First, the conventions are so strong that the companies might be blindsided by dramatic environmental change. This is a fair point. But we would argue that in general the excellent company values almost always stress being close to the customer or are otherwise externally focused. Intense

---

* The converse, apparently, is also true. When we were working for our first client in Japan on a problem that had nothing to do with organization, we happened to witness a major reorganization in process at the same time as our study. We were startled by the dramatic nature of the change and the speed with which it took place. Within a week, nearly all the top 500 executives had changed jobs, many had moved from Tokyo to Osaka or vice versa, the dust had settled, and business was proceeding as usual. We concluded that the Japanese were able to reorganize as seemingly ruthlessly as they did because security was always present; not security of position, for many were demoted or transfered to subsidiary companies, but security that had its roots in solid cultural ground and shared meanings.

customer focus leads the prototypical excellent company to be unusually sensitive to the environment and thus *more* able to adapt than its competitors.

For us, the more worrisome part of a strong culture is the ever present possibility of abuse. One of the needs filled by the strong excellent company cultures is the need most of us have for security. We will surrender a great deal to institutions that give us a sense of meaning and, through it, a sense of security. Unfortunately, in seeking security, most people seem all too willing to yield to authority, and in providing meaning through rigidly held beliefs, others are all too willing to exert power. Two frightening experiments, those of Stanley Milgram at Yale and Philip Zimbardo at Stanford, warn us of the danger that lurks in the darker side of our nature.

The first, familiar to many, are Stanley Milgram's experiments on obedience. Milgram brought adult subjects off the street into a Yale lab and asked them to participate in experiments in which they were to administer electric shocks to victims. (In fact, they were not doing so. The victims were Milgram conspirators and the electric shock devices were bogus. Moreover, the experimental protocol made it appear that the choice of both the victim and the shocker was random.) Initially, Milgram had the victims placed in one room and the shock givers in another. Following instructions given to them by a white-coated experimenter (the authority figure), the shock givers turned the dial, which went from "mild" to "extremely dangerous." On instruction, they administered the electricity, and to Milgram's surprise and disappointment, the experiment "failed." All went "all the way" in administering shock. One hundred percent followed orders, although in earlier written tests over 90 percent predicted they would not administer any shock whatsoever.

Milgram added embellishments. He connected the rooms with a window, so the shock givers could see the "victims" writhe in pain. He added victim "screams." Still, 80 percent went to "intense" on the dial, and 65 percent went to "extremely dangerous." Next he made the victims appear to be "homely, 40-year-old female accountants." He took the experiments out of the university and conducted them in a dreary downtown loft. He had the shock giver hold the victim's hand on the electric charge plate. All these steps were aimed at breaking down the subject's acceptance of the white-coated experimenter's authority. None worked very well. People still by and large accepted authority.

Milgram postulated numerous reasons for the outcome. Was it genetic? That is, is there species-survival value in hierarchy and authority that leads us all to submit? Are people simply sadistic? He concluded, most generally, that our culture "has failed almost entirely in inculcating internal controls on actions that have their origin in authority."

In the other case, Zimbardo advertised in a newspaper in Palo Alto, California (a prototypical upper-class community), soliciting volunteers for a "prison" experiment. At dawn one Saturday morning he went out, picked the volunteers up, booked them, and took them to a wallboard "prison" in the basement of the Stanford University psychology building. Within hours of their arrival, the randomly assigned "guards" started acting like guards and the randomly assigned "prisoners" started acting like prisoners. Well within the first twenty-four hours, the guards were behaving brutally—both physically and psychologically. By the end of the second day, a couple of the prisoners were on the verge of psychotic breakdown and had to be released from the experiment. "Warden" Zimbardo, afraid of his own behavior as well as that of the others, stopped the experiment four days into a ten-day protocol.

The lessons are applicable to the cultures of the excellent companies, but the apparent saving grace of the latter is that theirs are not inwardly focused. The world of the excellent company is especially open to customers, who in turn inject a sense of balance and proportion into an otherwise possibly claustrophobic environment.*

On the whole, we stand in awe of the cultures that the excellent

---

* Another worrisome aspect of the strong corporate culture is how well those who have spent most of their lives in it will fare on the outside should they ever leave, which some do. Our observation, though not backed by research, is that they do less well than might be expected, given their often stellar records in the top companies. It's a bit like a baseball pitcher traded away from the Yankees. These people often are totally unaware of the enormous support system they had going for them in the excellent company, and are at the very least initially lost and bewildered without it.

companies have built. Despite their inherent dangers, these cultures
have made their companies unique contributors to society. Grand
old Ma Bell, beleaguered though she currently may be by deregulation, gave America a telephone system that by almost any measure
is the best in the world. Theodore Vail's seventy-five-year-old insistence that the company was not a telephone company but a "service" company had everything to do with that achievement.

Finally, and paradoxically, the excellent companies appear to
take advantage of yet another very human need—the need one has
to control one's destiny. At the same time that we are almost too
willing to yield to institutions that give us meaning and thus a sense
of security, we also want self-determination. With equal vehemence, *we simultaneously seek self-determination and security.*
This is certainly irrational. Yet those who don't somehow learn to
manage the tension are, in fact, technically insane. In *Denial of
Death,* Ernest Becker stated the paradox: "Man thus has the absolute tension of the dualism. Individuation means that the human
creature has to oppose itself to the rest of nature [stick out]. Yet it
creates precisely the isolation that one can't stand—and yet needs
in order to develop distinctively. It creates the difference that becomes such a burden; it accents the smallness of oneself and the
sticking-outness at the same time."

Psychologists study the need for self-determination in a field
called "illusion of control." Stated simply, its findings indicate that
if people think they have even modest personal control over their
destinies, they will persist at tasks. They will do better at them.
They will become more committed to them. Now, one of the most
active areas of this experimentation is the study of cognitive biases.
The typical experiment here has subjects estimate their probability
of success at future tasks after they have had some experience doing the same sort of activity. The results are pretty consistent:
whether the subjects are adults or college sophomores, they overestimate the odds of succeeding at an easy task and underestimate
the odds of succeeding at a hard one. In short, they regularly dis-
tort estimates of the possibilities of events. If their proven past record is, say, 60 percent success at the easy task, the subjects will

likely estimate their future odds of success at 90 percent. If past demonstrated ability at the hard task is 30 percent, the subject will put 10 percent odds on success in the future. We need to succeed and stick out—desperately—so we overestimate the possibility of doing the easy task. And to preserve face and ensure security, we underestimate the possibility of getting the difficult task done.

A set of experiments that really highlights our need for self-determination and at the same time our desire for control is the "shut off the noise button" variety mentioned in the Introduction. Even though we never use the button, the fact that we could if we wanted to improves our performance by quantum steps. Other similar experiments produce similar results. A subject allowed to dip his own hand into the lottery bowl will believe the odds of drawing the winning ticket to be substantially higher than if someone else does the drawing. If a subject is given four cans of unmarked soft drinks to taste and then asked to choose his favorite, he will like his first choice much better than if the choice had been restricted to only two cans. (The drinks are the same beverage in all cases.) The fact, again, that we *think* we have a *bit* more discretion leads to *much* greater commitment.

And here, too, the excellent companies seem to understand these important, if paradoxical, human needs. Even in situations in which industry economics seem strongly to favor consolidation, we see the excellent companies dividing things up and pushing authority far down the line. These companies provide the opportunity to stick out, yet combine it with a philosophy and system of beliefs (e.g., Dana's overriding belief in "the productive people") that provide the transcending meaning—a wonderful combination.

TRANSFORMING LEADERSHIP

We often argue that the excellent companies are the way they are because they are organized to obtain extraordinary effort from ordinary human beings. It is hard to imagine that billion-dollar companies are populated with people much different from the norm for the population as a whole. But there is one area in which the excellent companies have been truly blessed with unusual leadership, especially in the early days of the company.

Leadership is many things. It is patient, usually boring coalition building. It is the purposeful seeding of cabals that one hopes will result in the appropriate ferment in the bowels of the organization. It is meticulously shifting the attention of the institution through the mundane language of management systems. It is altering agendas so that new priorities get enough attention. It is being visible when things are going awry, and invisible when they are working well. It's building a loyal team at the top that speaks more or less with one voice. It's listening carefully much of the time, frequently speaking with encouragement, and reinforcing words with believable action. It's being tough when necessary, and it's the occasional naked use of power—or the "subtle accumulation of nuances, a hundred things done a little better," as Henry Kissinger once put it. Most of these actions are what the political scientist James MacGregor Burns in his book *Leadership* calls "transactional leadership." They are the necessary activities of the leader that take up most of his or her day.

But Burns has posited another, less frequently occurring form of leadership, something which he calls "transforming leadership"—leadership that builds on man's need for meaning, leadership that creates institutional purpose. We are fairly sure that the culture of almost every excellent company that seems now to be meeting the needs of "irrational man," as described in this chapter, can be traced to transforming leadership somewhere in its history. While the cultures of these companies seem today to be so robust that the need for transforming leadership is not a continuing one, we doubt such cultures ever would have developed as they did without that kind of leadership somewhere in the past, most often when they were relatively small.

The transforming leader is concerned with minutiae, as well. But he is concerned with a different kind of minutiae; he is concerned with the tricks of the pedagogue, the mentor, the linguist—the more successfully to become the value shaper, the exemplar, the maker of meanings. His job is much tougher than that of the transactional leader, for he is the true artist, the true pathfinder. After all, he is both calling forth and exemplifying the urge for transcendence that unites us all. At the same time, he exhibits almost boorish consistency over long periods of time in support of his one or two transcending values. No opportunity is too small, no forum too insignificant, no audience too junior.

Burns speaks most convincingly of the leader's need to enable his followers to transcend daily affairs. He begins by faulting earlier students of leadership for their preoccupation with power, suggesting that such attention blinded them to the far more important task of instilling purpose. "This absolutely central value [purpose] has been inadequately recognized in most theories," he maintains. "Leadership over human beings is exercised when persons with certain motives and purposes mobilize, in competition or conflict with others, institutional, political, psychological and other resources so as to arouse, engage and satisfy the motives of followers." In essence, Burns says, "Leadership, unlike naked power wielding, is thus inseparable from followers' needs and goals." He thereby sets the stage for a concise definition of transforming leadership:

[Transforming leadership] occurs when one or more persons *engage* with others in such a way that leaders and followers raise one another to higher levels of motivation and morality. Their purposes, which might have started out separate but related, in the case of transactional leadership, become fused. Power bases are linked not as counterweights but as mutual support for common purpose. Various names are used for such leadership: elevating, mobilizing, inspiring, exalting, uplifting, exhorting, evangelizing. The relationship can be moralistic, of course. But transforming leadership ultimately becomes *moral* in that it raises the level of human conduct and ethical aspiration of both the leader and the led, and thus has a transforming effect on both. . . . Transforming leadership is dynamic leadership in the sense that the leaders throw themselves into a relationship with followers who will feel "elevated" by it and often become more active themselves, thereby creating new cadres of leaders.

Burns, like others, believes that leaders are appealing to certain unconscious needs: "The fundamental process is an elusive one; it is, in large part, *to make conscious what lies unconscious among followers.*" Taking Chairman Mao Tse-tung as exemplar, he comments: "His true genius was in understanding the emotions of others." The business psychologist Abraham Zaleznick makes much the same point in contrasting leaders and managers: "Managers prefer working with people; leaders stir emotion." The work of the psychologist David McClelland, notably in *Power: The Inner Experience,* provides an experimentally based description of the process:

[We] set out to find exactly, by experiment what kinds of thoughts the members of an audience had when exposed to a charismatic leader. . . . They were apparently strengthened and uplifted by the experience; they felt more powerful, rather than less powerful or submissive. This suggests that the traditional way of explaining the influence of a leader on his followers has not been entirely correct. He does not force them to submit and follow him by the sheer overwhelming magic of his personality and persuasive powers. . . . In fact, he is influential by strengthening and inspiriting his audience. . . . The leader arouses confidence in his followers. The followers feel better able to accomplish whatever goals he and they share.

Picking up on one of Burns's main points, leader-followers symbiosis, we find two attributes of that symbiosis especially striking: believability and excitement. On the first count, believability, we find that our value-infused top-performing companies are led by those who grew up with the core of the business—electric engineering at HP or Maytag, mechanical engineering at Fluor or Bechtel. The star performers are seldom led by accountants or lawyers. On the second count, excitement, Howard Head, inventor and entrepreneur, father of the Head ski and the Prince tennis racket, exhorts: "You have to believe in the impossible." At Hewlett-Packard, top management's explicit criterion for picking managers is their ability to engender excitement.

A simple description of the process of finding excitement is provided by James Brian Quinn, who is, among other things, a long-term student of the real, sloppy process of finding and achieving overarching strategic values and objectives. Quinn quotes a consumer goods chief executive officer: "We have slowly discovered that our most effective goal is *to be best* at certain things. We now try to get our people to help us work out what these things should be, how to define *best* objectively, and how to *become* best in our selected spheres. You would be surprised at how motivating that can be."

Warren Bennis has a good metaphor for the transforming leader—the leader as "social architect." But, to give credit where credit is due, Bennis, Burns, and we, in our comments on the excellent companies, were anticipated decades ago by both Chester Barnard, whom we'll meet again in the next chapter, and Philip Selznick, who published in 1957 an often-overlooked thin blue volume entitled *Leadership and Administration,* in which he says:

The inbuilding of purpose is a challenge to creativity because it involves transforming men and groups from neutral, technical units into participants who have a particular stamp, sensitivity, and commitment. This is ultimately an educational process. It has been well said that the effective leader must know the meaning and master the technique of the educator. . . . The art of the creative leader is the art of institution building, the reworking of human and technological materials to fashion an organism that embodies new and enduring values. . . . To institutionalize is to *infuse with value* beyond the technical requirements of the task at hand. The prizing of social machinery beyond its technical role is largely a reflection of the unique way it fulfills personal or group needs. Whenever individuals become attached to an organization or a way of doing things as persons rather than as technicians, the result is a prizing of the device for its own sake. From the standpoint of the committed person, the organization is changed from an expendable tool into a valued source of personal satisfaction. . . . The institutional leader, then, *is primarily an expert in the promotion and protection of values.*

We should pause briefly here, as we exalt values, to ask what values? Maybe, for one, we might suggest simply "to be best" in any area, as James Brian Quinn says, or to "be true to our own aesthetic," as Walter Hoving said of himself and Tiffany's. Perhaps it's Ray Kroc of McDonald's seeing "beauty in a hamburger bun," or Watson's "respect for the individual" at IBM, or Dana's belief in "the productive people" or "Forty-eight-hour parts service anywhere in the world" at Caterpillar. Corny? Only if we are cynical. Such values are transforming for the companies that live them.

Much of our discussion has verged on the high-sounding, for example, the talk of creating a transforming purpose. It *is* high-sounding, but at the same time it is simply practical. We have argued that man is quite strikingly irrational. He reasons by stories, assumes himself to be in the top 10 percent judged by any good trait, needs to stick out and find meaning simultaneously, and so on. Yet management practice seldom takes these foibles and limitations into account.

The excellent company managements, however, do take these things into account—either consciously or unconsciously. The result is better relative performance, a higher level of contribution from the "average" man. More significant, both for society and for the companies, these institutions create environments in which people can blossom, develop self-esteem, and otherwise be excited participants in the business and society as a whole. Meanwhile, the much larger group of nonexcellent performers seems to act, almost perversely, at odds with every variable we have described here. Losing instead of winning is the norm, as are negative rather than positive reinforcement, guidance by the rule book rather than tapestries of myths, constraint and control rather than soaring meaning and a chance to sally forth, and political rather than moral leadership.

PART THREE

BACK TO BASICS

4

Managing Ambiguity and Paradox

The test of a first-rate intelligence is the ability to hold two opposed ideas in mind at the same time and still retain the ability to function.

—F. Scott Fitzgerald

Some of the managers who have reviewed our eight attributes of management excellence comment that the attributes are interesting, but not necessarily basic—not crucial in explaining why the excellent companies perform so well. We think these managers are mistaken. Many otherwise smart and business-savvy people are operating from a theoretical base that is simply out of date. That is certainly understandable, for none of the new theory, whether right or wrong, is easily accessible. It is, after all, in a fairly early and messy stage of development. And it is largely obscure, tied to the "real world" only by implication, as is typically the case with leading-edge theory.

Thus, to understand the relationship between the performance of the excellent companies and the eight attributes, we need some new theory. And that is what we intend to provide. In this chapter we will try to combine some recent contributions in the evolution of management theory with some of the theoretical implications of the excellent companies data.

But let's return momentarily to the world of the rational model. The old management theories were attractive because they were straightforward and not laden with ambiguity or paradox. On the other hand, the world isn't like that. (Interestingly, one of our Japanese colleagues was highly critical of a report we had prepared for one of his clients. He said that it was too pat. He felt his clients would doubt the accuracy of anything so unambiguous.) We find it fascinating that the world of science is proceeding in paradoxical directions hauntingly similar to those we are observing and hypothesizing for the world of management theory. For example, light was first thought of as particles. Then scientists found that light behaved like waves. However, as soon as the wave view was adopted, new evidence for the particle theory poured in. But if light really is particulate, it ought to have mass, and then it couldn't travel at the speed of light, which of course it does. Heisenberg showed that you could know either the position of a subatomic particle or its mass, but not both at the same time. So here we see the most rational of disciplines, physics, rushing headlong through the looking glass of ambiguity, with atomic physicists using terms like "charm," "strangeness," "anti-matter," and "quark" to describe particles.

Science is easier to comprehend when we can grasp principles through metaphors from the world we know—things we have touched, seen, or smelled. Hence the appeal of Niels Bohr's model of the atom; it looked just like the solar system, a nucleus with electrons whirling around it like planets circling the sun. But unfortunately that view didn't get us very far in understanding the atom because the atom, we now know, is not much like the solar system. Similarly, the world of management seemed easier when we drew parallels with the military, most people's metaphor still for management structure in the twentieth century. But, again, the parallels broke down when we tried to understand anything more complex than, for instance, a regiment under fire. (There are arguably problems even in that unambiguous imagery. William Manchester, in *Good-bye, Darkness,* tells of Marine vets laughing derisively at the untested zeal—and orders—of young OCS lieutenants who would lead them into withering enemy fire. Many a young officer ended up going over the wall alone, and not coming back. So, as any seasoned hand well knows, the crystal-clear so-called military model—give an order and get instant compliance—doesn't even hold for the military.) We need something better if we are really to understand. Unfortunately, better isn't easier at first blush, although it may turn out to be easier as understanding improves. As we shall see, the new wave of management thought leads us to an ambiguous, paradoxical world—just like the world of science. But we think its tenets are more useful and ultimately more practical. Most important, we think the excellent companies, if they know any one thing, know how to manage paradox.

Numerous schemes have been elaborated to describe the evolution of management theories. For our purposes, the most useful starting place is one put forward by Richard Scott of Stanford. Scott imagines four main eras of both theoretical development and management practice. Each era is defined by the unique combination of elements in a two-dimensional grid. To picture it, think of one side as running from "closed" to "open" and the other side as running from "rational" to "social." Now, let's look at the first side of the spectrum, from closed to open. It proceeds from mechanical thinking about organizations (closed) to gestalt thinking (open). In marked contrast to the prevailing wisdom today, management theorists of the first sixty years of this century did not worry about the environment, competition, the marketplace, or anything else external to the organization. They had a "closed system" view of the world. That view, myopic as it now seems, centered on what ought to be done to optimize resource application by taking into account only what went on inside a company. It didn't really change much until almost 1960, when theorists began to acknowledge that internal organization dynamics were shaped by external events. Explicitly taking account of the effects of external forces on the organization's internal workings, then, launched the "open system" era.

The second side of Scott's grid runs from "rational" to "social." Rational, in this context, means that clear purposes and objectives for organizations exist, and that these can be determined rather straightforwardly. For example, if your company is in the mining business, your goal ought to be to maximize earnings from present mines and future exploration activities. If we take these purposes and objective functions as givens, top management has merely to choose the means that most efficiently result in the achievement of the goals. Rational decisions can be made on this basis, and the organization's course will be so charted. Social, on the other hand, acknowledges the messiness of determining purposes and implies that selection of purpose is not very straightforward or deductive. (For example, in our hypothetical mining business, just what is meant by "maximize"? How do we measure "earnings"—do we limit ourselves to hard rock mining only—and how do we make concrete decisions against anything so intangible as exploration success?) The social view supposes that decisions about objectives are value choices, not mechanical ones. Such choices are made not so much by clear-headed thinking as by social coalition, past habit patterns, and other dynamics that affect people working in groups.

The four distinct eras emerge when the two axes are juxtaposed (see figure on opposite page). The first runs from 1900 to around 1930, and is the "closed system–rational actor" era. The two main proponents of that era's theoretical position were Max Weber and Frederick Taylor. Weber was a German sociologist. He postulated the view that bureacracy—order by rule—is the most efficient form of human organization. Taylor, an American, put Weber's theories to the test with time and motion studies. The thrust of the Weber-Taylor school was to suggest that if a finite body of rules and techniques could be learned and mastered—rules about breakdown of work, about maximum spans of control, about matching authority and responsibility—then the essential problems of managing large groups of people would be more or less solved.

Weber and Taylor's dream, of course, was not realized, and the closed system–rational actor era was supplanted, from 1930 to 1960, by a closed system–social actor era. Its luminaries were Elton Mayo, Douglas McGregor, Chester Barnard, and Philip Selznick.

Mayo was a clinical psychologist working at the Harvard Business School who is mainly remembered as the father of the famous Hawthorne experiments. These investigations started out inauspiciously, as ordinary field work, consistent in most respects with the Taylor tradition. They were intended to be just a bunch of straightforward studies of industrial hygiene factors. The experiments took

FOUR STAGES OF THEORY AND LEADING THEORISTS

	Closed system	Open system
Rational actor	I. 1900-1930 Weber Taylor	III. 1960-1970 Chandler Lawrence Lorsch
Social actor	II. 1930-1960 Mayo et al. McGregor Barnard Selznick	IV. 1970-? Weick March

place chiefly in the bank wiring rooms of the Western Electric plant at Circero, Illinois, and were aimed at testing the effect of work conditions on productivity.

But a surprising series of events intruded on the theoretical background and continued to persist as stubbornly as the stubbornly held beliefs that preceded them. A good example is the one about lighting levels that we mentioned earlier: lights were turned up; productivity went up; lights were turned down; productivity went up again. What happened? The experiments were continued for a decade with continuing disconcerting results. Although the body of experimental data is so rich that many interpretations were and still are being made, the main point seems to be that the simple act of paying positive attention to people has a great deal to do with productivity. This effect permeates data on our excellent companies. Hewlett-Packard values innovation from a large body of workers, and its systems for paying attention to innovation (e.g., talking about it, honoring it) are crystal clear on that score. Managements of mining companies that are good in exploration have scores of ways of paying attention to field geologists.

Mayo and his followers at Harvard established the field of industrial social psychology. World War II spurred the field's growth, as it did so many others', and by the end of the war such related fields as group training and leadership selection were beginning to flower. After the war, a major contribution was made by Douglas McGregor. We chiefly remember McGregor for his development of Theory X and Theory Y, the opposing views that workers are lazy and need to be driven and, alternatively, that they are creative and should be given responsibility. McGregor's thesis was sweeping, as he noted in the preface to his landmark book *The Human Side of Enterprise:* "This volume is an attempt to substantiate the thesis that the human side of enterprise is 'all of a piece'—that the theoretical assumptions management holds about controlling its human resources determine the whole character of the enterprise." McGregor railed against the rationalist approach of the Taylor school. "If there is a single assumption that pervades conventional organization theory," he stormed, "it is that authority is the central, indispensable means of managerial control." McGregor noted that in reality authority exists as one of several forms of social influence and control, but sadly both the literature and the practicing managers of the time regarded authority as an absolute, rather than a relative, concept.

McGregor termed Theory X "the assumption of the mediocrity of the masses." Its premises are "(1) that the average human has an inherent dislike of work and will avoid it if he can, (2) that people, therefore, need to be coerced, controlled, directed, and threatened with punishment to get them to put forward adequate effort toward the organization's ends and (3) that the typical human prefers to be directed, wants to avoid responsibility, has relatively little ambition, and wants security above all." McGregor argued that Theory X is not a straw man, "but is in fact a theory which materially influences managerial strategy in a wide sector of American industry."

Theory Y, by contrast, assumes:

(1) that the expenditure of physical and mental effort in work is as natural as in play or rest—the typical human doesn't inherently dislike work; (2) external control and threat of punishment are not the only means for bringing about effort toward a company's ends; (3) commitment to objectives is a function of the rewards associated with their achievement—the most important of such rewards is the satisfaction of ego and can be the direct product of effort directed toward an organization's purposes; (4) the average human being learns, under the right conditions, not only to accept but to seek responsibility; and (5) *the capacity to exercise a relatively high degree of imagination, ingenuity, and creativity in the solution of organizational problems is widely, not narrowly, distributed in the population.* [Italics ours]

McGregor's theories and those that followed, in what was to become the "human relations" school of management, have fallen into disrepute over the last decade. The overwhelming failure of the human relations movement was precisely its failure to be seen as a balance to the excesses of the rational model, a failure ordained by its own equally silly excesses. We are reminded of one company that went off the deep end on T-groups, bottom-up planning, democratic management, and other forms of a "make everyone happy" work environment. The positive result was that if Jean was smoking in a meeting and that bothered Joe, Joe learned to be very comfortable about asking Jean to stop, and Jean learned not to take that request personally. The company was apparently very adroit in solving that bugaboo of all large companies, better communications. The problem was that, although it really was good at communications on the little things, somehow the big issues never got raised.

Whereas the rational model was a pure top-down play, the social model, as produced by McGregor's misguided disciples, became a pure bottom-up play, an attempt to start revolutions via the training department. McGregor had feared that all along and said, "The assumptions of Theory Y do not deny the appropriateness of authority, but they do deny that it is appropriate for all purposes and under all circumstances."

We are beginning to perceive, however dimly, a central theme that in our minds makes the excellent companies great. On the surface of it, Theory X and Theory Y are mutually exclusive. You pick one or the other. As a leader, you are authoritarian or you are democratic. In reality, you are neither and both at the same time. Messrs. Watson (IBM), Kroc (McDonald's), Marriott, *et al.*, have been pathbreakers in treating people as adults, in inducing practical innovation and contributions from tens of thousands, in providing training and development opportunities for all, in treating all as members of the family. Mr. Watson, in fact, in carrying out his open door policy, had an unfailing weakness for the worker; his managers rarely won when a worker complaint was surfaced. On the other hand, all of these gentlemen were tough as nails. All were ruthless when their core values of service to the customer and unstinting quality were violated. They combined, then, a caring side and a tough side. Like good parents, they cared a lot—and expected a lot. To oversimplify their characteristics as predominantly "X-ish" or "Y-ish" is almost entirely to miss the point.

While McGregor and Mayo epitomize the social theory of organization applied to the individual human being, Chester Barnard and Philip Selznick, starting at about the same time as the other two, may yet emerge as the more influential theorists. In our view, the work of Barnard and Selznick has been grossly neglected by practicing managers.

Barnard, after having been president of New Jersey Bell, retired to Harvard to ponder his experience, writing *The Functions of the Executive* in 1938. Its density makes it virtually unreadable; nonetheless it is a monument. Harvard's Kenneth Andrews, in his introduction to a thirtieth-anniversary edition of the book (1968), said: "Barnard's aim is ambitious. As he tells us in his own preface, his purpose is first to provide a comprehensive theory of cooperative behavior in formal organizations. Cooperation originates in the need of an individual to accomplish purposes to which he is by himself biologically unequal."

While Mayo, McGregor, and others, including Barnard himself, were developing ideas aimed at calling forth the best efforts of people down the line, it was Barnard alone who sensed the unconventional and critical role of executives in making it all happen. In particular, Barnard concluded that it is the executive who must secure commitment and actively manage the informal organization. And he must do this while ensuring that the organization simultaneously achieves its economic goals. Barnard's was probably the first balanced treatment of the management process.

Barnard was also the first (we know of) to talk about the primary role of the chief executive as the shaper and manager of shared values in an organization: "The essential functions [of the executive] are, first, to provide the system of communications; second, to promote the securing of essential efforts; and third, to formulate and define purpose." He added that organizational values and purpose are defined more by what executives do than by what they say. "It has already been made clear that, strictly speaking, purpose is defined more nearly by the aggregate of action taken than by any formulation in words." He also emphasized that purpose, to be effective, must be accepted by all the contributors to the system of efforts. In the excellent companies, we see just that. Values are clear; they are acted out minute by minute and decade by decade by the top brass; and they are well understood deep in the companies' ranks.

Perhaps Barnard's genius is best expressed by his unusual stress on managing the whole:

The common sense of the whole is not obvious, and in fact is often not effectively present. Control is dominated by a particular aspect—the economic, the political, the religious, the scientific, the technological—with the result that [top performance] is not secured and failure ensues or is perpetually threatened. No doubt the development of a crisis due to unbalanced treatment of all the factors is the occasion for corrective action on the part of executives who possess the art of sensing the whole. A formal and orderly conception of the whole is rarely present, perhaps even rarely possible, except to a few men of executive genius, or a few executive organizations the personnel of which are comprehensively sensitive and well integrated.

Today it is still unusual to find emphasis on managing the whole.

Little more than a decade after Barnard's book appeared, Philip Selznick unveiled a similar theory in which he invented such terms as "distinctive competence" (what a particular company is uniquely good at, and most others are not) and "organizational character" (in which he anticipates the idea of organizations as cultures). We quote Selznick at length because we think he beautifully describes organizational character, competence, institutional values, and leadership. We find the traits, as he describes them, to be basic to the success of the excellent companies:

The term "organization" thus suggests a certain bareness, a lean, no-nonsense system of consciously coordinated activities. It refers to an *expendable tool,* a rational instrument engineered to do a job. An "institution," on the other hand, is more nearly a natural product of social needs and pressures—a responsive, adaptive organism. ... The terms institution, organizational character, and distinctive competence all refer to the same basic process—the transformation of an engineered, technical arrangement of building blocks into a social organism. ... Organizations become institutions as they are infused with values. . . . The infusion produces a distinct identity. Where institutionalization is well advanced, distinctive outlooks, habits and other commitments are unified, coloring all aspects of organizational life and lending it a *social integration* that goes well beyond formal coordination and command. . . . It is easy to agree to the abstract proposition that the function of an executive is to find a happy joinder of means and ends. It is harder to take that idea seriously. There is a strong tendency in administrative life to divorce means and ends by overemphasis on one or the other. The cult of efficiency in administrative practice is a modern way of overstressing means in two ways . . . by fixing attention on maintaining the smooth-running machine or by stressing the techniques of organization. . . . Efficiency as an operating ideal presumes goals are settled and resources available. In many situations, including most of the important ones, goals may not have been defined [or] when they are defined, the means necessary may still have to be created. Creation of means is not a narrow technical matter, it involves molding the social character of the institution. Leadership goes beyond efficiency (1) when it sets basic mission and (2) when it creates a social organism capable of fulfilling that mission.

The Mayo–McGregor–Barnard–Selznick legacy, the legacy of man as social actor, is immense. Unfortunately, as we have noted, the first two were discredited when naive disciples perverted their ideas, and the second two have never been, to this day, widely read or acclaimed. In particular, two of our findings (the correlatives of autonomy and entrepreneurship and of productivity through people) are wholly consistent with McGregor; three others (hands-on, value-driven; stick to the knitting; and simultaneous loose-tight properties) are of a piece with Barnard and Selznick's view of things. But still something is missing. We return to Scott's formulation.

Stage three, which lasted from about 1960 to 1970, was at once a step backward and a step forward. Scott calls it the "open system–rational actor" era. Theory took a step backward in that it reverted to mechanistic assumptions about man. It took a step forward in that the theorists were finally viewing a company as part of a competitive marketplace, shaped and molded by forces outside itself. A seminal contribution to the era was made by Alfred Chandler in *Strategy and Structure*. Quite simply, Chandler observed that organizational structures in great companies like Du Pont, Sears, General Motors, and General Electric are all driven by changing pressures in the marketplace. For example, Chandler traces the market-driven proliferation of product lines in both Du Pont and General Motors. He shows how that proliferation led to a needed shift away from a functional monolithic organizational form toward a more loosely coupled divisional structural form.

Chandler did this work at Harvard, and two other Harvard professors, Paul Lawrence and Jay Lorsch, followed it in 1967 with another landmark study, *Organization and Environment*. Their model was substantially more sophisticated than Chandler's, but reached roughly the same conclusion. They looked at organization structures and management systems and contrasted the top performers in a fast-moving business—specialty plastics—with the top performers in a stable, slow-moving business—containers. They found that the stars in the business characterized by stability maintained a simple functional organization form and simple control systems. By contrast, the stars in the fast-moving specialty plastics businesses had a more decentralized form and richer systems than their competitors who were not doing as well.

Finally, Scott postulates a fourth epoch, starting about 1970 and continuing to the present. He describes its theoretical position as "open system–social actor." Messiness dominates in both dimensions. The rational actor is superseded by the complex social actor, a human being with inbuilt strengths, weaknesses, limitations, contradictions, and irrationalities. The business insulated from the outside world is superseded by the business buffeted by a fast-paced, ever-changing array of external forces. In the view of today's leading theorists, everything is in flux—ends, means, and the storms of external change. The leaders of this era include Cornell's Karl Weick and Stanford's James March.

The dominant paradigm in this fourth epoch of organizational thought emphasizes informality, individual entrepreneurship, and evolution. The clearest signal that the leading management thinkers are radically departing from past views is the shift in metaphors. Weick is vehement on the subject of metaphor shift, contending that the usual military metaphors severely limit our ability to think about management sensibly: "Organizations have staff, line, and a chain of command. They develop strategy and tactics. Organizations attack competitors, recruit MBAs ... They solve problems by discharging people (honorably or otherwise), tightening controls, introducing discipline, sending for reinforcements, or clarifying responsibilities—since that's what you do when an army sags." Weick is convinced that military metaphors are a bad choice when it comes to the problem of managing a commercial enterprise. First, the use of military metaphor assumes that someone clearly wins and someone else clearly loses. In business, this is usually not the case. Second, Weick argues that the military metaphor is a bad choice because people solve problems by analogy, and as long as they use the military analogue, "It forces people to entertain a very limited set of solutions to solve any problem and a very limited set of ways to organize themselves."

The new metaphors, per Weick and March, which do open up rich new veins for thinking about managing—however threatening they may be to executives steeped in the old school—include sailing, playfulness, foolishness, seesaws, space stations, garbage cans, marketplaces, and savage tribes. As we discuss the excellent companies, we will suggest many others, such as champions, skunk works, and czars, which come from the ways the excellent companies talk about themselves. "Diverse as they are," argues Weick, "each metaphor has articulated some property of organizations that might otherwise have gone unnoticed." As Anthony Athos puts it, "the truth *lurks* in the metaphor."

Chester Barnard wrote *The Functions of the Executive* in 1938; it probably deserves to be called a complete management theory. So does Herbert Simon's *Administrative Behavior,* written in 1947. March and Simon's joint work *Organizations,* written in 1958, includes 450 interrelated propositions about organizing. It, too, constitutes a full management theory.

Arguably, there has been no true organizing theory written since then. Maybe March would contend that his book *Ambiguity and Choice in Organizations,* co-written with Johan Olsen in 1976, is a full-blown theory, but we think not. Certainly Karl Weick does not contend that his marvelous *Social Psychology of Organizing* is a fully developed theory. In fact, he says simply, "This book is about organizational appreciation."

The point is that the efforts by today's leading theorists add up to an important set of vignettes about managing. In crucial ways, these vignettes accurately contravene much of the conventional wisdom that existed previously. What is more, they counter old shibboleths in ways that are entirely congenial with our observations about excellent companies. But that is not to say that there is no need for new theory. The need is desperate if today's managers, their advisers, and the teachers of tomorrow's managers in the business schools are to be up to the challenges we posed in Chapter 2.

Certainly we are not proposing a complete theory of organizing here. But we do think that via the excellent companies findings we see a few dimensions of theory that have not been given attention by scholars or practicing managers. Moreover, we think that these findings provide us with a simple and direct way to express some concepts hitherto obscured in today's state-of-the-art theories. Meanwhile, there are a few underlying ideas that ought to be brought out as a basis, at least, for understanding the eight attributes we will be discussing in the next eight chapters.

The clear starting point is acceptance of the limits of rationality, the central theme of the last two chapters. Building on that, four prime elements of new theory would include our observations on basic human needs in organizations: (1) people's need for meaning; (2) people's need for a modicum of control; (3) people's need for positive reinforcement, to think of themselves as winners in some sense; and (4) the degrees to which actions and behaviors shape attitudes and beliefs rather than vice versa.

There are some very important ideas from past and current management theory that need to be woven into the fabric of new theory. Two that we particularly want to stress, because we don't think they have received the attention they deserve, are (1) the notion of companies, especially the excellent ones, as distinctive cultures; and (2) the emergence of the successful company through purposeful, but specifically unpredictable, evolution.

THE IMPORTANCE OF CULTURE

Some colleagues who have heard us expound on the importance of values and distinctive cultures have said in effect, "That's swell, but isn't it a luxury? Doesn't the business have to make money first?" The answer is that, of course, a business has to be fiscally sound. And the excellent companies are among the most fiscally sound of all. But their value set *integrates* the notions of economic health, serving customers, and making meanings down the line. As one executive said to us, "Profit is like health. You need it, and the more the better. But it's not why you exist." Moreover, in a piece of research that preceded this work, we found that companies whose only articulated goals were financial did not do nearly as well financially as companies that had broader sets of values.

Yet it's surprising how little is said about the shaping of values in current management theories—particularly how little is said about companies as cultures. The estimate of 3M quoted in Chapter 1— "The brainwashed members of an extremist political sect are no more conformist in their central beliefs"—remember, is the same 3M that's known not for its rigidity but for its unbridled entrepreneurship. Delta Airlines lives its "Family Feeling," and, notes chairman Tom Beebe, "What Delta has going for it is the very close relationship we all feel for one another." Some people leave Texas Instruments because it is "too rigid"; on the other hand, it has been tremendously innovative, and chairman Mark Shepherd says of its Objectives, Strategies, and Tactics planning system,"OST would be sterile were it not for the culture of innovation that permeates the institution." A *Fortune* analyst makes the following observation about Maytag: "The reliability of Maytag washers owes a lot to the Iowa work ethic." Columbia University's Stanley Davis claims, "Firms operating out of Rochester, New York [e.g., Kodak], or Midland, Michigan [e.g., Dow], often have very strong corporate cultures. Much stronger than firms that operate out of New York City or Los Angeles."

A few audible murmurings about values and culture have been made by the academics since Barnard and Selznick raised the issue. Richard Normann, in *Management and Statesmanship,* talks of the importance of the "dominating business idea," and comments that the "most crucial process" going on in any company may be the continuing interpretation of historic events and adjustment of the dominating business idea in that context. And in a recent book on organizational structuring, Henry Mintzberg mentions culture as a design principle, but only briefly, calling it (unfortunately) the "missionary configuration" and giving it a regrettable futuristic slant: "The missionary [structural] configuration would have its own prime coordinating mechanism—socialization, or, if you like, the standardization of norms—and a corresponding main design parameter—indoctrination. . . . The organization would have . . . an ideology. The perceptive visitor would 'sense it' immediately." But there's nothing as futuristic about it as Mintzberg implies. Procter & Gamble has been operating that way for about 150 years, IBM for almost 75. Levi Strauss's predominantly people-oriented philosophy started with an unheard-of "no layoff" policy following the 1906 San Francisco earthquake.

Andrew Pettigrew sees the process of shaping culture as the prime management role: "The [leader] not only creates the rational and tangible aspects of organisations, such as structure and technology, but also is the creator of symbols, ideologies, language, beliefs, rituals, and myths." Using strikingly similar language, Joanne Martin of Stanford thinks of organizations as "systems composed of ideas, the meaning of which must be managed." Martin has spurred a great deal of practical, specific research that indicates the degree to which rich networks of legends and parables of all sorts pervade top-performing institutions. HP, IBM, and DEC are three of her favorite examples. The research also indicates that the poor performers are relatively barren in this dimension. Warren Bennis also speaks of the primacy of image and metaphor:

It is not so much the articulation of goals about what an [institution] *should* be doing that creates new practice. It's the imagery that creates the understanding, the compelling moral necessity that the new way is right.... It was the beautiful writing of Darwin about his travels on the *Beagle,* rather than the content of his writing, that made the difference. Because the evolutionary idea had really been in the air for a while. Not only were there parallel mentions of it, but Darwin's uncle had done some of the primary work on it.... Thus, if I were to give off-the-cuff advice to anyone trying to institute change, I would say, "How clear is the metaphor? How is that understood? How much energy are you devoting to it?"

The business press, starting sometime in 1980, has increasingly used culture as a metaphor of its own. *Business Week* legitimated the practice by running a cover story on corporate culture in the late summer of 1980. Now the word seems to pop up more and more frequently in business journalism.

Perhaps culture was taboo as a topic following William H.Whyte, Jr.'s *The Organization Man* and the conformist, gray flannel suit image that he put forward. But what seems to have been overlooked by Whyte, and management theorists until recently, is what, in Chapter 12, we call the "loose-tight" properties of the excellent companies. In the very same institutions in which culture is so dominant, the highest levels of true autonomy occur. The culture regulates rigorously the few variables that do count, and it provides meaning. But within those qualitative values (and in almost *all* other dimensions), people are encouraged to stick out, to innovate. Thus, "IBM Means Service" underscores the company's overpowering devotion to the individual customer; but that very formulation also provides remarkable space. Everyone, from clerks on up, is prodded to do whatever he or she can think of to ensure that the individual customer gets taken care of. In a more mundane setting, Steven Rothman, writing in *D&B Reports,* quotes a Tupperware dealer: "The company gives me great freedom to develop my own approach. There are certain elements that need to be in every party to make it successful, but if those elements are colored by

you, a Tupperware dealer—purple, pink and polka dot, and I prefer
it lavender and lace—that's okay. That freedom allows you to be
the best you are capable of being." So, in fact, the power of the
value is in large measure that it encourages practical innovation to
carry out its spirit to the full.

EVOLUTION

To the extent that culture and shared values are important in unifying the social dimensions of an organization, managed evolution is
important in keeping a company adaptive.

We are confronted with an extraordinary conundrum. Most current theory is neither tight enough nor loose enough. Theory is not
tight enough to consider the role of rigidly shared values and culture as the prime source of purpose and stability. It proposes rules
and goal setting to cover these bases. At the same time, most current theory is not loose enough to consider the relative lack of structure and the need for wholly new management logic to ensure continuous adaptation in large enterprises. Instead, it habitually
proposes structural rules and planning exercises—both forms of rigidity—to hurdle this need.

Both problems proceed from the inherent complexity of large organizations, yet both have been banished by the excellent companies on an ad hoc basis. Big institutions are too complex, really, to
manage by rule books, so managers, to simplify the problem, use a
few transcending values covering core purposes. Adaptation is also
too complex to manage by rules in a big enterprise, so astute managers simply make sure that enough "blind variations" (i.e., good
tries, successful or not) are going on to satisfy the laws of probability—to ensure lots of bunt singles, an occasional double, and a once-
a-decade home run.

We need new language. We need to consider adding terms to our
management vocabulary: a few might be temporary structures, ad
hoc groups, fluid organizations, small is beautiful, incrementalism,
experimentation, action orientation, imitations, lots of tries, unjustified variations, internal competition, playfulness, the technology of

foolishness, product champions, bootlegging, skunk works, cabals, and shadow organizations. Each of these turns the tables on conventional wisdom. Each implies both the absence of clear direction and the simultaneous need for action. More important still, we need new metaphors and models to stitch these terms together into a sensible, coherent, memorable whole.

James March, as we noted, has proposed as a concomitant to his "garbage can" metaphor a model of decision making in which "streams of problems, solutions, participants, and choice opportunities" swirl around, occasionally resulting in decisions. Moreover, he suggests that "[we] need to supplement the technology of reason with a technology of foolishness. Individuals and organizations need ways of doing things for which they have no good reason. Not always. Not usually. But sometimes. They need to act before they think." Leadership in such a system, March asserts, would play a different role: "Rather than an analyst looking for specific data, we are inclined to think of a monitor looking for unusual signals." March sums up his views more attractively when he notes that "such a vision of managing organizations is a relatively subtle one. It assumes that organizations are to be sailed rather than driven, and that the effectiveness of leadership often depends on being able to time small interventions so that the force of natural organizational processes amplifies the interventions rather than dampens them." And in his loveliest image of all, he says that "organizational design is more like locating a snow fence to deflect the drifting snow than like building a snowman."

Karl Weick chooses to describe adaptation in terms of "loosely coupled systems." He argues that most management technology has wrongly assumed tight coupling—give an order or declare a policy, and it is automatically followed. "The more one delves into the subtleties of organizations," says Weick, "the more one begins to question what order means and the more convinced one becomes that prevailing preconceptions of order (that which is efficient, planned, predictable, and survived) are suspect as criteria for successful evolution." He suggests that two evolutionary processes are at the heart of adaptation. "Unjustified variation is critical," he states, adding, "I am most sympathetic to purposeful complication." Next, he urges that "retrospective sense making is the key metaphor." By that he means that management's prime task is to select, after the fact, from among "experiments" naturally going on in the organization. Those that succeed and are in accord with management's purposes are labeled after the fact ("retrospective sense making") as harbingers of the new strategic direction. The losers are victims of trying to learn from "impoverished, shallow surroundings." That is, there's little to learn from; the company is marked by few "good tries." Weick logically concludes: "No one is ever free to do something he can't think of." And he provides a description by Gordon Siu of a marvelous experiment to clinch his point:

. . . If you place in a bottle half a dozen bees and the same number of flies, and lay the bottle down horizontally, with its base to the window, you will find that the bees will persist, till they die of exhaustion or hunger, in their endeavor to discover an issue through the glass; while the flies, in less than two minutes, will all have sallied forth through the neck on the opposite side. . . . It is their [the bees] love of light, it is their very intelligence, that is their undoing in this experiment. They evidently imagine that the issue from every prison must be there where the light shines clearest; and they act in accordance, and persist in too logical action. To them glass is a supernatural mystery they never have met in nature; they have had no experience of this suddenly impenetrable atmosphere; and, the greater their intelligence, the more inadmissible, more incomprehensible, will the strange obstacle appear. Whereas the feather-brained flies, careless of logic as of the enigma of crystal, disregarding the call of the light, flutter wildly hither and thither, and meeting here the good fortune that often waits on the simple, who find salvation there where the wiser will perish, necessarily end by discovering the friendly opening that restores their liberty to them.

Weick concludes:

This episode speaks of experimentation, persistence, trial and error, risks, improvisation, the one best way, detours, confusion, rigidity, and randomness all in the service of coping with change. Among the most striking contrasts are those between tightness and looseness. There are differences in the degree to which means are tied to ends, actions are controlled by intentions, solutions are guided by imitation of one's neighbor, feedback controls search, prior acts determine subsequent acts, past experience constrains present activity, logic dominates exploration, and in the degree to which wisdom and intelligence affect coping behavior. In this example loose ties provide the means for some actors to cope successfully with a serious change in their environment. Each individual fly, being loosely tied to its neighbor and its own past, makes numerous idiosyncratic adaptations that eventually solve the problem of escape. Looseness is an asset in this particular instance but precisely how and when looseness contributes to successful change and how change interventions must be modified to cope with the reality of looseness is not obvious.

Weick, March, and others are fascinated by the role that classic evolutionary processes play in the development of organizations. Their role in linking the populations of companies to the needs of the environment has always been recognized by economists: if companies do not stay fit and relevant, they do not survive. In the broadest sense (albeit a very disconcerting one for most managements), the theory works all too well. Most of today's *Fortune* 500 were not there fifty years ago. All of the private sector's net new jobs in the United States during the past twenty years were added by companies not on the *Fortune* 1000 twenty years ago; two thirds of the net new jobs came from companies with fewer than twenty employees twenty years ago. Ten years ago our automobile giants seemed invincible. Today we wonder whether more than one will survive.

In 1960, Theodore Levitt of Harvard wrote an article in the *Harvard Business Review,* "Marketing Myopia," in which he pointed out that every industry was once a growth industry. Perversely, a vicious cycle sets in. After experiencing continued growth for a while, managers in the industry come to believe that continuing growth is assured. They persuade themselves that there is no competitive substitute for their product, and develop too much faith in the benefits of mass production and the inevitable steady cost reduction that results as output rises. Managements become preoccupied with products that lend themselves to carefully controlled improvement and the benefits of manufacturing cost reduction. All of these forces combine to produce an inevitable stagnation or decline.

In *Dynamic Economics,* the economist Burton Klein puts forward a carefully researched and very similar view: "Assuming that an industry has already reached the stage of slow history, the advances will seldom come from the major firms in the industry. In fact, of some fifty inventions [fifty key twentieth-century breakthrough innovations that he studied] that resulted in new S-shaped curves [major new growth patterns] in relatively static industries, I could find no case in which the advance in question came from a major firm in the industry." George Gilder elaborates on Klein's work: "The very process by which a firm becomes most productive in an industry tends to render it less flexible and inventive."

It appears that evolution is continuously at work in the marketplace; that adaptation is crucial; and that few big businesses, if any, pull it off. Many of our excellent companies most probably will not stay buoyant forever. We would merely argue that they've had a long run—a much longer and more successful run than most—and are coming much closer than the rest to maintaining adaptability and size at the same time.

We believe that one major reason for this, only recently of concern to the management theorists, is intentionally seeded evolution within the companies. The excellent companies are *learning organizations.* They don't wait around for the marketplace eventually to do them in; they create their own internal marketplace. (One analyst noted that IBM's real management magic in the days of 90 percent market share was creating, almost from whole cloth, the specter of competitors.) Intriguingly, the top companies have developed a whole host of devices and management routines to stave off calcification. They experiment more, encourage more tries, and permit small failures; they keep things small; they interact with customers—especially sophisticated customers—more (all functions of the organization); they encourage internal competition and allow resultant duplication and overlap; and they maintain a rich informal environment, heavily laden with information, which spurs diffusion of ideas that work. Interestingly, few are very articulate about what they're up to. The best, at HP, 3M, Digital, Wang, J&J, or Bloomingdale's, are especially inarticulate about the role of management in orchestrating such a process. They know it when they see it and can detect deterioration at the margin; but they, like we, have no sound language with which to describe the phenomenon. Patrick Haggerty came as close as anyone with his OST system at TI to institutionalizing innovation. Yet even here, because of its orderly, systems-like nature, TI is showing unfortunate signs of regularly suppressing rather than encouraging continued adaptation.

A decade ago, Peter Drucker anticipated the need for adaptation when he said, in *The Age of Discontinuity,* "Businessmen will have to learn to build and manage innovative organizations." Norman Macrae, deputy editor of *The Economist,* hinted that "constant reorganization is the main reason why I judge that big American corporations are still often the most efficient day-to-day business operators in the world." Igor Ansoff, a long-time student of business strategy, adds: " . . . we can predict the loss of primacy of structure as the leading component in defining organizational capability. Structure will become a dynamic enabler of both change and unchange, the ultimate model of 'organized chaos.' " We're reminded of an analysis we conducted of successful versus unsuccessful mineral exploration departments in major mining companies. As we reported to the client, all the successful explorers looked to us like "nothing so much as structured chaos." "Buzzing, blooming environments" is the way our colleague David Anderson aptly characterized the excellent companies in a very early report on this research.

It seems to add up to a small is beautiful, small is effective philosophy among the companies that are good. Repeatedly we found things a lot more divided up and a lot less tidy than they should be according to conventional wisdom. Again, what's going on? Whatever happened to economies of scale? How can these companies be cost-effective? Don't they understand learning curve economics? In a section entitled "It Seemed Like a Good Idea at the Time," *Science 82* reported:

Ten years ago Ford Motor Co. built a plant to produce 500,000 tons of iron engine blocks a year. Erected on the principle that mass production means lower costs, it was four stories high and large enough to enclose 72 football fields. But the plant designed to produce V-8 engines turned out to be too big and too specialized. When new designs for lighter engines followed the oil crunch, Ford discovered that retooling the huge plant was prohibitively expensive. It shut down the factory, moving operations to a 30-year-old, smaller plant.

The excellent companies understand that beyond a certain surprisingly small size, *diseconomies* of scale seem to set in with a vengeance. In early 1980, when we reported on our tentative results to John Doyle, vice president of research and development at HP, we commented that the top-performing companies we had talked to, including HP, seemed to be "suboptimizing" their divisions and plants (regularly making them smaller than either market factors or economies of scale would seem to dictate). Although we meant it as a favorable comment, he took great exception to our choice of words. "For us, what you're calling 'suboptimal' is optimal," he maintained with vehemence.

Throughout the remaining chapters of this book we will encounter examples of things that are not organized to be as tidy as the rule books prescribe. The common theme, the thread that seems to tie the apparent untidiness together, is the idea that small is effective. We found divisions, plants, and branches that were smaller than any cost analysis would suggest they should be. We found "simulated entrepreneurship"; here Dana's "store managers" (plant managers, in fact) are a good example. Decentralization of function was practiced where classic economics would ordain otherwise; that is, Dana's approximately ninety store managers can all have their own cost-accounting system, each do his own purchasing, each control virtually all aspects of personnel policy. In company after company, we found ten-person skunk works that were regularly more innovative than fully equipped R&D and engineering groups with casts of hundreds. We found example after example of internal competition, of various teams working on the same thing, of product-line duplication and overlap, of people experimenting and pointing with pride to their useful mistakes. We found myriads of tiny, quick-hit task forces, more quality circles than American managements were supposed to be using at the time. We observed less standardization of procedure and a concomitant greater willingness to "let them do it any way they want if it makes sense and works."

We believe we are breaking some important theoretical ground here. We observed more "chunking," more breaking things up into manageable units than others professedly have. In theory to date, the small is effective idea is usually limited to discussions of innovativeness by small firms. In most of the excellent companies, however, we see various approaches to chunking as a main tenet of effective management practice. Interestingly, the more we look at the phenomenon, the more we see it as a vehicle for enhanced efficiency as well as a vehicle to foster adaptation and survival.

Oliver Williamson at the University of Pennsylvania is the leading theoretician on the efficiency front. His book *Markets and Hierarchies* probably hasn't gotten the attention it deserves because it is so difficult to read (even the author admits as much in the Preface). Williamson argues that, in conventional estimates of scale economies, we have vastly underestimated "transaction costs," which means the cost of communication, coordination, and deciding. It is roughly the same point we made earlier in connection with the geometrical increase in complexity associated with arithmetic growth in numbers of employees, if they need to interact to get tasks done. To the extent that many factors need to be coordinated, the costs of coordination usually swamp technologically determined economies of scale. Williamson's assertions are supported by a growing body of empirical evidence.

Williamson's ideas are close to what we have observed, but with a vital difference. He sees the world as either black or white. If the transaction costs indicate that a function might better be performed efficiently by markets (e.g., outsiders) than by hierarchies, then it

must be contracted out. To use a trivial example to illustrate, in a
big professional office, watering plants seems a minor chore. Yet
deciding on which greenery fits the season and keeping it alive
turns out to absorb a good deal of staff time. Therefore, it works
out to be less expensive (and more effective) to contract with an
outside plant provisionary and watering service. (The creator of the
service is usually a clever entrepreneur who realizes what a pain in
the neck plant maintenance often becomes.) If things can be done
more efficiently inside, then he argues that hierarchies are the
norm. We believe that the market option is fully available inside
the company. The core management practices of IBM, HP, 3M,
TI, McDonald's, Delta, Frito, Tupperware, Fluor, J&J, Digital,
and Bloomingdale's bear strongly on the point that markets of all
kinds work well inside. Internal competition has been a formally
mandated policy at P&G since 1930; Sloan explicitly used it at GM
beginning in the early twenties.

Tidiness is sacrificed and efficiency is gained. In fact, more than
efficiency is gained. Through chunking, a corporation encourages a
high volume of rapid action. The organization *acts,* and then learns
from what it has done. It experiments, it makes mistakes, it finds
unanticipated success—and new strategic direction inexorably
emerges. We strongly believe that the major reason big companies
stop innovating is their dependence on big factories, smooth production flow, integrated operations, big-bet technology planning, and
rigid strategic direction setting. They forget how to learn and they
quit tolerating mistakes. The company forgets what made it successful in the first place, which was usually a culture that encouraged action, experiments, repeated tries.

Indeed, we believe that the truly adaptive organization evolves in
a very Darwinian way. The company is trying lots of things, experimenting, making the right sorts of mistakes; that is to say, it is
fostering its own mutations. The adaptive corporation has learned
quickly to kill off the dumb mutations and invest heavily in the ones
that work. Our guess is that some of the most creative directions
taken by the adaptive organizations are not planned with much precision. These organizations are building March's snow fences to de-

flect the tries, experiments, mistakes, and occasional grand successes in directions that are only roughly right. Our colleague Lee Walton argues, in fact, that management's principal job is "to get the herd heading roughly west."

A primary criticism of our use of the Darwinian analogy is that it appears to limit itself to small, incremental innovations. Big breakthroughs, like IBM's System 360, these critics say, require surehanded, "bet the company" planning. We like to have the question raised, because it is so easy to refute on both theoretical and empirical grounds. There seems to be no support in evolutionary theory for a narrow incremental interpretation (i.e., that evolution proceeds by tiny steps). The evolutionary biologist Stephen Jay Gould, an undisputed leader in the field, points out that the evolution of the human brain in a random variation, for instance, far from occurring as a tiny or logical next-step advance for the species, was 50,000 or more years ahead of its time; that is, it provided gross overcapacity for caveman's needs. For that reason, it has not basically changed since then. Of course, major successful mutations are much rarer than small ones. But that, surely, is what we would expect. In any event, the evolutionary model does support the occurrence of big leaps without requiring, in Gould's words, an all-knowing God or prescient planning.

The empirical evidence is even more striking. Burton Klein and others have demonstrated in scores of studies that, in industry, it is *never* the industry leader who makes the big leap. On the contrary, they claim, it is the inventor or small guy who makes the big leap, even in stodgy industries like steel and aluminum in which one wouldn't expect to find many inventors around. Moreover, our own investigations indicate hardly less—that most of the big new business breakthroughs, from McDonald's (breakfast menu items pulling in about 40 percent of the business) to GE (engineered plastics and aircraft engines) have come from small bands of zealots operating outside the mainstream. Indeed, a long-time observer noted that *no* major IBM product introduction in the last quarter century has come from the formal system. That's not to say that the company doesn't place a big, well-planned bet on a new product or business at some point. Of course it does. It is to say that the mutation itself, even the big one, occurs pretty far down the line and invariably under the tutelage of outside-the-system zealots. Even more support is added by the fact that almost no big innovation (so labeled after the fact) is ever used as originally intended. As we've said, computers were seen to have only a handful of applications, many of them at the Census Bureau. Transistors were developed for a tiny set of military uses. Diesel locomotives originally were perceived as useful only in freight yard switching. Xerography was aimed at a small, existing part of the lithography market; mass copies were not at all a driving force in either the invention or the early marketing.

So the evolutionary, somewhat untidy theory of management holds for large- as well as small-scale innovations, and for efficiency as well as effectiveness. One final element of the theory deserves prominent mention. In biology, isolation can spell disaster in an active species zone. Mutations (the equivalent of new product tries) may occasionally occur, but selections (successes) are unlikely. Thus, the process of mutant generation (experiments, tries, mistakes) must deal not with isolation but with real business needs and opportunities. The excellent company solution is that it occurs via a remarkably rich set of interactions with the environment—namely, customers. Here again, conventional theory falls woefully short of excellent company reality.

Management theory took a major turn about fifteen years ago. As we noted, the environment finally seeped into models of organizing. The landmark study was done in 1967 by Lawrence and Lorsch. More recently, the two leading proponents of evolutionary theory have been two star young researchers, Jeffrey Pfeffer and Gerald Salancik. In 1978 they published *The External Control of Organizations: A Resource Dependence Perspective.* Also in 1978 Marshall Meyer published *Environments and Organizations,* which included seven theoretical chapters and a recapitulation of about ten major, decade-long research programs. All these researchers' hearts are in the right place. For example, take Pfeffer and Salancik: "The central thesis of this book is that to understand the behavior of an organization, you must understand the context of that behavior. Organizations are inescapably bound up with the conditions of the environment. Indeed, it has been said that all organizations engage in activities which have as their logical conclusion adjustment to the environment." There's nothing wrong with that. We find it intriguing, however, that in reviewing the indexes of these three clearly bellwether works, we could not unearth the word "customer," or "client" or "clientele." All three books talk about the environment, but wholly miss the excellent company richness of customer contact, which encompasses scores of devices from subway interviews under Bloomingdale's store in New York (largely symbolic) to vast arrays of user experiments at Digital and elsewhere.

A few researchers have gone farther. In particular, James Utterback and Eric von Hippel at MIT, studying higher technology companies, have done several analyses of the intensity of customer contact among the better-performing companies. Utterback, for instance, talks about the outreach of innovative firms: "It implies special connections with your environment, not general connections. And connections with particularly creative and demanding users. And it demands that the connection be informal and personal. . . . A lot of translating and testing goes on between the producer of technology and the customer. Often there is a great deal of interaction between the possible users and the organization that brings a major product change into the market." But Utterback and von Hippel's writings are not mainstream, and are limited in scope to a relatively small population of high technology companies. The phenomenon of intense company-customer linkage that we observed, we are pleased to say, knew no industry boundaries.

There's nothing new under the sun. Selznick and Barnard talked about culture and value shaping forty years ago. Herbert Simon begain talking about limits to rationality at the same time. Chandler began writing about environmental linkages thirty years ago. Weick began writing about evolutionary analogues fifteen years ago. The problem is, first, that none of the ideas has yet become mainstream; they have had little or no effect on practicing businessmen. Second, and we think more important, all of them fall far, far short of depicting the richness, the variety of linkages that we observed in the excellent companies. It's not just experimenting; it's thousands of experiments that characterize these operations. It's not just internal competition; it's doing virtually all resource allocation by internal competition. It's not just small is beautiful; it's hundreds of very small units, a tiny fraction of the technologically attainable size. It's not just customer contact, but a vast array of devices for getting everyone from the junior accountant to the CEO in regular contact with the customer. In short, the core management practices in the excellent companies aren't just different. They set conventional management wisdom on its ear.

5

A Bias for Action

Eighty percent of success is showing up.
—Woody Allen

But above all try something.

—FDR

Ready. Fire. Aim.

—Executive at Cadbury's

There's an excitement about being in the game parks of East Africa that's impossible to describe. The books don't do it. The slides and movies don't do it. The trophies most of all don't do it. When you're there, you feel it. People who've been there can hold one another in rapt conversation for hours about it; people who haven't been there can't quite imagine it.

We experience some of the same helplessness in describing an excellent company attribute that seems to underpin the rest: action orientation, a bias for getting things done. For example, we were trying to depict to an executive responsible for project management coordination how it might be possible radically to simplify the forms, procedures, paperwork, and interlocking directorates of committees that had overrun his system. We said, quite off-handedly, "Well, at 3M and TI they don't seem to have these problems. People simply talk to each other on a regular basis." He looked at us blankly. Our words hardly sounded like exotic advice—or even helpful advice. So we said, "You're not competing with 3M. Let's go to St. Paul for a day and take a look. You'll be surprised."

Our friends at 3M were tolerant of the excursion, and we observed all sorts of strange goings-on. There were a score or more casual meetings in progress with salespeople, marketing people, manufacturing people, engineering people, R&D people—even accounting people—sitting around, chattering about new-product problems. We happened in on a session where a 3M customer had come to talk informally with about fifteen people from four divisions on how better to serve his company. None of it seemed rehearsed. We didn't see a single structured presentation. It went on all day—people meeting in a seemingly random way to get things done. By the end of the day our friend agreed that our description had been fairly accurate. Now *his* problem was the same as ours: he didn't know how to describe the situation to anyone else.

It's very difficult to be articulate about an action bias, but it's very important to try, because it is a complex world. Most of the institutions that we spend time with are ensnared in massive reports that have been massaged by various staffs and sometimes, quite literally, hundreds of staffers. All the life is pressed out of the ideas; only an iota of personal accountability remains. Big companies seem to foster huge laboratory operations that produce papers and patents by the ton, but rarely new products. These companies are besieged by vast interlocking sets of committees and task forces that drive out creativity and block action. Work is governed by an absence of realism, spawned by staffs of people who haven't made or sold, tried, tasted, or sometimes even seen the product, but instead, have learned about it from reading dry reports produced by other staffers.

However, life in most of the excellent companies is dramatically different. Yes, they too have task forces, for example. But one is more apt to see a swarm of task forces that last five days, have a few members, and result in line operators' doing something differently rather than the thirty-five-person task force that lasts eighteen months and produces a 500-page report.

The problem we're addressing in this chapter is the all-too-reasonable and rational response to complexity in big companies: coordinate things, study them, form committees, ask for more data (or new information systems). Indeed, when the world is complex, as it is in big companies, a complex system often does seem in order. But this process is usually greatly overdone. Complexity causes the lethargy and inertia that make too many companies unresponsive.

The important lesson from the excellent companies is that life doesn't have to be that way. The excellent companies seem to abound in distinctly individual techniques that counter the normal tendency toward comformity and inertia. Their mechanism comprises a wide range of action devices, especially in the area of management systems, organizational fluidity, and experiments—devices that simplify their systems and foster a restless organizational stance by clarifying which numbers really count or arbitrarily limiting the length of the goal list.

ORGANIZATIONAL FLUIDITY: MBWA

Both Warren Bennis in *The Temporary Society* and Alvin Toffler in *Future Shock* identified the need for the adhocracy as a way of corporate life. In rapidly changing times, they argued, the bureaucracy is not enough. By "the bureaucracy," they mean the formal organization structure that has been established to deal with the routine, day-in, day-out items of business—sales, manufacturing, and so on. By "the adhocracy," they mean organizational mechanisms that deal with all the new issues that either fall between bureaucratic cracks or span so many levels in the bureaucracy that it's not clear who should do what; consequently, nobody does anything.

The concept of organizational fluidity, therefore, is not new. What *is* new is that the excellent companies seem to know how to make good use of it. Whether it's their rich ways of communicating informally or their special ways of using ad hoc devices, such as task forces, the excellent companies get quick action just because their organizations are fluid.

The nature and uses of communication in the excellent companies are remarkably different from those of their nonexcellent peers. The excellent companies are a vast network of informal, open communications. The patterns and intensity cultivate the right people's getting into contact with each other, regularly, and the chaotic/anarchic properties of the system are kept well under control simply because of the regularity of contact and its nature (e.g., peer versus peer in quasi-competitive situations).

The intensity of communications is unmistakable in the excellent companies. It usually starts with an insistence on informality. At Walt Disney Productions, for instance, everyone from the president on down wears a name tag with only his or her first name on it. HP is equally emphatic about first names. Then come the open door policies. IBM devotes a tremendous amount of time and energy to them. The open door was a vital part of the original Watson philosophy, and it remains in force today—with 350,000 employees. The chairman continues to answer all complaints that come in to him from any employee. Open door use is pervasive at Delta Airlines as well; at Levi Strauss it means so much that they call the open door the "fifth freedom."

Getting management out of the office is another contributor to informal exchanges. At United Airlines, Ed Carlson labeled it "Visible Management" and "MBWA—Management By Walking About." HP treats MBWA ("Management By Wandering Around" in this instance) as a major tenet of the all-important "HP Way."

Another vital spur to informal communication is the deployment of simple physical configurations. Corning Glass installed escalators (rather than elevators) in its new engineering building to increase the chance of face-to-face contact. 3M sponsors clubs for any groups of a dozen or so employees for the sole purpose of increasing the probability of stray problem-solving sessions at lunchtime and in general. A Citibank officer noted that in one department the age-old operations–versus–lending-officer split was solved when everybody in the group moved to the same floor with their desks intermingled.

What does it add up to? Lots of communication. All of HP's golden rules have to do with communicating more. Even the social and physical settings at HP foster it: you can't wander around long in the Palo Alto facilities without seeing lots of people sitting together in rooms with blackboards, working casually on problems. Any one of those ad hoc meetings is likely to include people from R&D, manufacturing, engineering, marketing, and sales. That's in marked contrast to most large companies we've worked with, where the managers and analysts never meet or talk to customers, never meet or talk to salesmen, and never look at or touch the product (and the word "never" is not chosen lightly). A friend at HP, talking about that company's central lab organization, adds: "We aren't really sure what structure is best. All we know for certain is that we start with a remarkably high degree of informal communication, which is the key. We have to preserve that at all costs." 3M's beliefs are similar, which led one of its executives to say, "There's only one thing wrong with your excellent company analysis. You need a ninth principle—communications. We just plain talk to each other a lot without a lot of paper or formal rigmarole." All of these examples add up to a virtual *technology of keeping in touch,* keeping in constant informal contact.

In general, we observe the tremendous power of the regular, positive peer review. A simple tale comes from Tupperware. Tupperware makes about $200 million in pre-tax earnings on about $800 million in sales of simple plastic bowls. The key management task is motivating the more than 80,000 salespeople, and a prime ingredient is "Rally." Every Monday night all the saleswomen attend a Rally for their distributorship. At Rally, everyone marches up on stage—in the reverse order of last week's sales—during a process known as Count Up (while their peers celebrate them by joining in All Rise). Almost everyone, if she's done anything at all, receives a pin or badge—or several pins and badges. Then they repeat the entire process with small units marching up. On the one hand, this is a fairly punishing drill—straight head-on-head competition that can't be avoided. On the other hand, it is cast with a positive tone: everybody wins; applause and hoopla surround the entire event; and the evaluation technique is informal rather than paper-laden. In fact, the entire Tupperware system is aimed at generating good news opportunities and celebration. Every week there is an array of new contests. Take any three moribund distributorships: management will give a prize to whichever one has the best sales increase in the next eight weeks. Then there are the thirty days of Jubilee each year in which *15,000* are feted (3,000 at a time in week-long events) with awards, prizes, and ceremonies of all kinds. The entire environment is one that utilizes, in the extreme, positive reinforcement.

Above all, when we look at HP, Tupperware, and others, we see a very conscious management effort to do two things: (1) honor with all sorts of positive reinforcement any valuable, completed action by people at the top and more especially way down the line; and (2) seek out a high volume of opportunities for good news swapping.

We should note that when we were doing the first round of survey interviews, the three principal interviewers gathered together after about six weeks. When we tried to summarize what seemed most important (and different) to us, we unanimously agreed that it was the marvelously informal environments of the excellent companies. We have not changed our view since. The name of the successful game is rich, informal communication. The astonishing by-product is the ability to have your cake and eat it, too; that is, rich informal communication leads to more action, more experiments, more learning, and simultaneously to the ability to stay better in touch and on top of things.

Now consider this. "The Chase senior executive's voice was tinged with reluctant admiration," reports *Euromoney.* "If they don't like it at Citibank, they change it—not gradually, like we would, but immediately, even if they have to turn the bank upside down to do it." And this: one IBM executive commented, "It is said that back in the 1960s, IBM set an objective of being able to mount a major reorganization in just a few weeks." IBM's values remain constant, and the attendant stability permits it structurally to shift major hunks of resources to attack a particular problem. At the smaller end of the spectrum, the CEO of successful TRAK, a $35 million sporting goods company, noted that in order to keep his stars turned on he had to move to a flexible organization: "You've got to keep coming up with new projects to hang on to valuable people . . . [Our approach] is flexible reorganization and task teams. We're making it a permanent part of our organizing scheme."

Again, Harris Corporation has done the virtually impossible: it has largely licked the problem of diffusing research funded by the government into areas that are commercially viable. Many others have tried, and almost all have failed. The prime ingredient in Harris's success is that the management regularly shifts chunks of engineers (twenty-five to fifty) out of government projects and moves them, as a group, into new commercial venture divisions. Similar moves have been crucial to Boeing's success. One officer notes: "We can do it [create a big new unit] in two weeks. We couldn't do it in two years at International Harvester."

There are scores of variations on this theme in the excellent companies, but they all come down to a refreshing willingness readily to shift resources: chunks of engineers, chunks of marketers, products among divisions, and the like.

Chunking

We vividly recall walking into the office of a top-flight line officer who was now a "product group coordinator." He was a tough old nut who had won his spurs solving labor negotiation problems. Now his desk was bare, and he thumbed idly through a *Harvard Business Review* collection of human relations articles. When we talked about what he was up to, he produced a list of committees that he chaired. This illustration adds up, de facto, to the matrix; it adds up to an environment of fragmented responsibilities. It does not add up to what we found in the excellent companies.

The line officer who has headed one of Exxon's Asian affiliates for the last ten years made a presentation on "strategy" at a recent top management meeting. He reported a remarkable tale of improvement. Was it a tale of shrewd foresight and bold strategic moves? Not in our view. It was a story, instead, of a series of pragmatic actions. In almost every one of the ten years, some single problem had been knocked off. One year a blitzkrieg group came through from regional headquarters and helped him get receivables under control. Another year, the attack was aimed at closing down some unprofitable segments. In another year, a further blitz effort helped work out a novel arrangement with distributors. It was a classic example of what we have come to call the "theory of chunks." We have come to believe that the key success factor in business is simply getting one's arms around almost any practical problem and knocking it off—now. Exxon in Japan simply executed (to near perfection) a series of practical maneuvers. They made each problem manageable. Then they blitzed it. The time associated with each program was fairly short. That it was the *real* number one priority for that short period of time was unquestioned. It sounded like strategic foresight, but we'd argue that it was a much more remarkable trait: they had just gotten a string of practical tasks done right.

There is an underlying principle here, an important trait of the action orientation that we call chunking. That simply means breaking things up to facilitate organizational fluidity and to encourage action. The action-oriented bits and pieces come under many labels—champions, teams, task forces, czars, project centers, skunk works, and quality circles—but they have one thing in common. They never show up on the formal organization chart and seldom in the corporate phone directory. They are nevertheless the most visible part of the adhocracy that keeps the company fluid.

The small group is the most visible of the chunking devices. Small groups are, quite simply, the basic organizational building blocks of excellent companies. Usually when we think of organizational building blocks, we focus on higher levels of agglomeration—departments, divisions, or strategic business units. Those are the ones that appear on the organization charts. But in our minds, the small group is critical to effective organizational functioning. In this sense (as well as many others) the excellent companies look very Japanese. In *Japan As Number One*, Ezra Vogel said that the entire business and social structure of Japanese companies is built around the Kacho (section head) and the eight- to ten-person group that typically comprises a section:

The essential building block of a company is not a man with a particular role assignment and his secretary and assistants. The essential building block of the organization is the section. . . . The lowly section, within its sphere, does not await executive orders but takes the initiatives. . . . For this system to work effectively leading section personnel need to know and to identify with company purposes to a higher degree than persons in an American firm. They achieve this through long experience and years of discussion with others at all levels.

Apparently the small group as building block works in the United States as well, although not as so natural a part of the national culture as it does in Japan. In the new-product area, 3M has several hundred four- to ten-person venture teams running about. Or recall TI's 9,000 teams zipping about looking for small productivity improvements. In Australia, one of the few large companies with an excellent labor record is ICI. Among the programs that managing director Dirk Ziedler implemented in the early 1970s was a series of interlocking teams that look very much like the Japanese section.

The true power of the small group lies in its flexibility. New-product teams are formed anywhere at 3M and nobody worries very much about whether or not they fit exactly into division boundaries. Appropriately, TI chairman Mark Shepherd calls his company "a fluid, project-oriented environment." The good news from the well-run companies is that what ought to work does work.

It's also quite remarkable how effective team use in the excellent companies meets, to a tee, the best academic findings about the makeup of effective small groups. For instance, the effective productivity or new product teams in the excellent companies usually range from five to ten in size. The academic evidence is clear on this: optimal group size, in most studies, is about seven. Other findings are supportive. Teams that consist of *volunteers,* are of *limited duration,* and *set their own goals* are usually found to be much more productive than those with the obverse traits.

The Ad Hoc Task Force. The task force can be the epitome of effective chunking. Unfortunately, it can also become the quintessence of hopeless bureaucracy. How well we remember the analysis! The client was a $600 million sector of a several-billion-dollar company. We inventoried the task forces and there were 325 of them formally in existence. So far, not much news. What really floored us, and the company in turn, was that not a single task force had completed its charge in the last three years. Not a single one had been disbanded either. In a similar situation with another client we randomly picked task force reports and found that the typical length was well over one hundred pages; signoffs ran from twenty on up to nearly fifty.

Let's quickly review recent history to understand the current love affair with task forces. Although they undoubtedly existed previously in many unlabeled forms, NASA and the Polaris program gave them a good name. NASA invented the ad hoc team structure and in early programs delivered the goods. The Polaris submarine program worked even better. The task force notion then diffused to industry and was used for everything. By 1970, it had become incorporated so pervasively in many large companies that it had become just one additional part of the rigid system it was meant to fix.

In hindsight, several things went wrong. Like any other tool adopted within a bureaucratic context, it eventually became an end in itself. Paper pushing and coordination took the place of task-directed activity. Stodgy, formal, paperbound, rule-driven institutions layered the task force on a maze that lay beneath, rather than using it as a separable, action-inducing chunk. Task forces became nothing more than coordinating committees—with a different name. Like other management tools adopted in the wrong context, the task force made things worse, not better.

That's the bad news. The good news is that in organizations in which the context is right—ready acceptance of fluidity and adhocracy—the task force has become a remarkably effective problem-solving tool. In effect, it is the number one defense against formal matrix structures. It acknowledges the need for multifunctional problem solving and implementation efforts, but not through the establishment of permanent devices.

A story helps to illustrate our point. In the midst of this survey, one of us walked into Digital headquarters in Maynard, Massachusetts, on a blustery February day. After we had finished the formal part of the interview, we asked one executive to describe some of the actual stuff he would be working on for the next few days. We wanted to get a flavor for the way things really work at Digital.

He said that he and six other people from the company were about to reorganize the national sales force. Each of the seven is a senior line manager. Each has full authority to sign off on the change for his group. We were talking to this fellow on a Thursday. He and the group would be leaving for Vail, Colorado (they're no fools at Digital), that evening. He said, "We'll be back by Monday night, and I expect we'll announce the changes in the sales force on Tuesday. The front end of the implementation should be well in place a week or so later."

As we did more interviewing, we repeatedly heard variations on this theme. The hallmarks of task force work that we found at such disparate companies as Digital, 3M, HP, TI, McDonald's, Dana, Emerson Electric, and Exxon were strikingly different from the bureaucratic model we had come to expect from so many other situations. At the excellent companies, task forces were working the way they are supposed to work.

There aren't many members on these task forces, usually ten or less. They really are the incarnation of the small group properties we talked about earlier. The unfortunate contrasting tendency in the bureaucratic model is to involve everyone who might have an interest. Task force membership typically balloons to the twenties, and we've even seen a few with as many as seventy-five members. The point is to limit active task force participation to the principal actors. That wouldn't work in many companies, because it requires trust on the part of those left out that they will be represented well.

The task force reporting level, and the seniority of its members, are proportional to the importance of the problem. If the problem is a big one, virtually all members are senior people and the task force reports to the chief executive. It is essential that the people have the charter to make stick whatever they recommend. A Digital executive said, "We want senior members only, no substitutes. The kinds of people we want are busy people whose main objective is to get off the damned task force and get back to work." We call this the "busy member theorem."

The duration of the typical task force is very limited. This is a compelling characteristic. At TI, it's rare if any task force lasts more than four months. Among the exemplary companies, the idea that any task force could last more than six months is repugnant.

Membership is usually voluntary. This was explained to us best at 3M: "Look, if Mike asks me to serve on a task force, I will. That's the way we do things. But it had better be a real problem. There'd better be some results. If there aren't, I'm damned if I'm going to waste my time helping Mike again. If it's my task force, I'll try to make sure that those who spend time on it get real value from it."

The task force is pulled together rapidly, when needed, and usually not accompanied by a formal chartering process. Since task force work is the primary means of problem solving in complex, multifunctional environments, the survey companies, fortunately, are able to pull them together at the drop of a hat and with little fanfare. By contrast, in the bureaucracy of 325 task forces described earlier, formal written charters (often lengthy) accompanied each task force.

Follow-up is swift. TI is exemplary in this regard. We are told that three months after a task force is formed, senior management wants to know what happened as a result. "Nothing; we're still working on a report," is not a satisfactory answer.

No staff are assigned. About half of the 325 task forces mentioned earlier had permanent staff assigned to them: paper shufflers associated with a paper-shuffling group. In *no* instance at TI, HP, 3M, Digital, or Emerson was there a report of a "staff" person permanently assigned to a task force as an executive director, an "assistant to," or a full-time report writer.

Documentation is informal at most, and often scant. As one executive told us, "Task forces around here are not in the business of producing paper. They are in the business of producing solutions."

Finally, we must reiterate the importance of context, of climate. The necessity of open communications was underscored by IBM's Frederick Brooks in his discussion of the System 360 development, for which he was a principal architect. Although this was a giant project team and much larger in scope than what is typically meant by a task force, the structure was fluid. According to Brooks, reorganizations took place with great regularity. Contact among members was intense; all principal players met in conference for a half day each week to review progress and decide on changes. Minutes were published within less than twelve *hours*. Everyone on the project had access to all the information he needed: every programmer, for instance, saw all the material coming from every group on the project. Nobody who attended the weekly meetings came in an advisory (i.e., staff) role. "Everyone had the authority to make binding commitments," says Brooks. The System 360 group had annual "supreme court" sessions, which typically lasted two full weeks. Any problems not solved elsewhere got resolved in this intensive two-week interchange. Most companies we have observed couldn't conceive of sending twenty key players off for two weeks; or of meeting together for a half day each week. Nor could they conceive of widespread information sharing or meetings at which all participants had the authority to make binding commitments.

The difference between this and the way so many other organizations do business is so striking that one more example from the nonexcellent side seems a fitting close to this section. We were recently asked to review why a computer-based management information system project was not working. This project crossed many organizational boundaries and had been organized as a task force. We pieced together a case history of its activities over the previous year, and found that, although it was following most of the rules of good task force management, the computer people and the division people were almost never in face-to-face communication, except at formal meetings. They could, for example, have moved into a common facility; a small group, they could even have worked in the same room. But neither was willing to do so. On trips to the field, they could have stayed in the same hotel, but they never did. One side claimed it was staying in less expensive hotels; the other countered that it was staying closer to the plants. They could have at least dined together after hours on field trips, but one side liked to play tennis and the other didn't. It all sounds pretty silly, and client executives didn't believe us, initially. But when we finally got all the people in the same room, they reluctantly agreed that we were right on every score. It would be nice to report that it got better after that, but it never did. The project, sound in all respects from a business standpoint, was eventually scrubbed.

Project Teams and Project Centers. The analysis of the task force is a favorite. Everyone does it, yet the excellent companies use this mundane tool quite differently from the rest. The task force is an exciting, fluid, ad hoc device in the excellent companies. It is virtually *the* way of solving and managing thorny problems, and an unparalleled spur to practical action.

IBM organized for the System 360 project by using the very large task force or project team, another form of adhocracy. People say that the project moved forward with lots of fits and starts, but the System 360's organization, particularly in its later years, clearly attracted the institution's top talent and set it to work on the monumental task—with no distractions. Companies like Boeing, Bechtel, and Fluor use massive project teams like this all the time. Indeed, it is fundamental to their way of doing business as so much of their business is project work. They have an impressive ability to shift rapidly between structures—their routine structure for day-to-day affairs and their project team structure. What is perhaps even more impressive, though, is to see a big company that *doesn't* routinely use project teams shifting into this mode with the ease of an experienced driver shifting gears. That seemed to be the case with IBM and the System 360, and we are impressed.

General Motors provides another particularly striking example of use of the temporary structure. The automotive industry is under attack. Virtually everything American automotive management does seems to be a day late and a dollar short. Yet we are impressed by any $60 billion institution that can beat its principal domestic competitors by almost three years on an implementation task, which is exactly what GM did with its downsizing project. The principal vehicle was the project center, a classic temporary organization. GM's project center took 1,200 key people out of the historically autonomous GM' divisions—including the division's most important people, like the chief engineers—and put them in the project center. The center lasted for four years. It had a clear task: to get the downsizing job fully specified, under way, and passed back to the divisions for final implementation. The real magic to the story is that when the task was accomplished, the project center for downsizing disappeared in 1978. GM, in fact, was so pleased with the downsizing success that it has chosen to adopt project centers as a prime mode of organizing for the eighties. Eight project centers now exist in a special project center building. Two of these are currently working on the electric car and overall engine computerization; another is working on labor issues.

Most organizations, when confronted with an overwhelming strategic problem, either give it to planning staffs or tack it onto the objectives of numerous otherwise busy line managers. If staff is supposed to solve the problem, commitment never develops. If the usual line organization is supposed to solve it, momentum never develops. IBM's System 360 or GM's downsizing project are dramatic, promising examples of the way in which problems like this can be successfully attacked.

The Japanese use this form of organization with frightening alacrity. To build a world-competitive position in, say, robotics or microcomputing, the Japanese pull key people from various companies into project centers to do the basics in development research. When the key technological problems have been solved, the key people go back to their own companies and compete like crazy with one another. Products are then ready for the world—after they've been honed by tough competition within Japan.

Honda's CVCC program is an example. Key people were pulled off all other tasks and put on the CVCC project for several years. Canon did the same thing in developing its Canon AE-1; the company bundled 200 of its senior engineers together in "Task Force

X" for two and a half years until the AE-1 was developed, implemented, and successfully launched in the marketplace.

There are numerous other examples of chunking, and we will come to them in later sections of the book. At the moment, however, there are four main messages that we want to get across about chunking. First, ideas about cost efficiency and economies of scale are leading us into building big bureaucracies that simply cannot act. Second, the excellent companies have found numerous ways (not just a few) to break things up in order to make their organizations fluid, and to put the right resources against problems. Third, all the chunking and other devices will not work unless the context is right. Attitudes, climate, and culture must treat ad hoc behavior as more normal than bureaucratic behavior. Finally, the free-wheeling environments in which ad hoc behavior flourishes are only superficially unstructured and chaotic. Underlying the absence of formality lie shared purposes, as well as an internal tension and a competitiveness that make these cultures as tough as nails.

EXPERIMENTING ORGANIZATIONS

"Do it, fix it, try it," is our favorite axiom. Karl Weick adds that "chaotic action is preferable to orderly inaction." "Don't just stand there, do something," is of the same ilk. Getting on with it, especially in the face of complexity, does simply come down to trying something: Learning and progress accrue only when there is *something* to learn from, and the something, the stuff of learning and progress, is any completed action. The process of managing this can best be thought of in terms of the experiment and, on a more pervasive basis, the experimenting process.

The most important and visible outcropping of the action bias in the excellent companies is their willingness to try things out, to experiment. There is absolutely no magic in the experiment. It is simply a tiny completed action, a manageable test that helps you learn something, just as in high-school chemistry. But our experience has been that most big institutions have forgotten how to test and learn. They seem to prefer analysis and debate to trying something out, and they are paralyzed by fear of failure, however small. The problem was accurately described recently in *Science*. NASA "invented" a technique called Success Oriented Management (SOM) to control space shuttle development. It assumes that everything will go right. As one official put it, "It means you design everything to cost and then pray." The intention was to eliminate parallel and possibly redundant development in test hardware, in response to the current cost pressures facing the agency. But as *Science*—and others—have noted, the program has led to wholesale deferrals of difficult work, embarrassing accidents, expensive redesigns, erratic staffing, and the illusion that everything is running well. "The net effect of this management approach," says *Science*, "has been an absence of realistic plans, inadequate understanding of the status of the program, and the accumulation of schedule and cost deficits without visibility."

Nowhere has the problem been more obvious than in the development of the space shuttle's three main engines. *Science* reports, "Rather than test each engine component separately, NASA's main contractor simply bolted it all together, and—with fingers crossed—turned on the power. At least five major fires resulted." Under the influence of SOM, NASA officials began to confuse prediction with reality (in fairness, this was probably forced on them by political reality). NASA suffered from "technological hubris," says a Senate analyst. "Managers became overconfident that technological breakthroughs would materialize to save the situation." This is certainly not the NASA of old, where redundancy was purposeful, testing took place regularly, and programs were on time— and worked.

The similarity and abundance of such tales is frightening, and they add up to nothing less than common management practice. For example, a giant bank prepared to introduce travelers' checks into a highly competitive market. A task force labored eighteen months and produced a cabinetful of market analyses. As the nationwide launch date approached, we asked the head of the project what he had done in the way of hard market testing. He answered that he had talked to two banker friends in Atlanta about carrying the checks. "Two?" was our incredulous reaction. "Two," he affirmed. "We weren't sure the project would be approved. We didn't want to tip our hand."

We hear feeble excuses like this day in and day out. On the other hand, we were impressed by an incisive comment made by a friend at Crown Zellerbach, a competitor of P&G in some paper product markets. "P&G tests and tests and tests. You can see them coming for months, often years. But you know that when they get there, it is probably time for you to move to another niche, not to be in their way. They leave no stone unturned, no variable untested." P&G is apparently not afraid of testing and therefore telegraphing its moves. Why? Because, we suspect, the value added from learning before the nationwide launch so far exceeds the costs of lost surprise.

Getting on with it marks P&G and most of the excellent companies. Charles Phipps, of Texas Instruments, describes the company's early success, its willingness to be bold and daring. He captures the spirit of the experiment—TI's ability to learn quickly, to get something (almost anything) out in the field. "They surprised themselves: as a very small company, $20 million, with very limited resources, they found they could outmaneuver large laboratories like Bell Labs, RCA and GE in the semiconductor area, because they'd just go out and try to *do* something with it, rather than keep it in the lab."

Example after example reflected the same experimenting mentality. At Bechtel, senior engineers talk about their guiding credo, maintaining a "fine feel for the doable." At Fluor, the principal success factor may be what they call "taking an idea and making metal out of it." At Activision, the watchword for video-game design is "build a game as quickly as you can. Get something to play with. Get your peers fooling with it right away. Good ideas don't count around here. We've got to see something." At a successful $25 million designer household goods operation, Taylor & Ng in San Francisco, owner Win Ng describes his philosophy: "Developing a prototype early is the number one goal for our designers, or anyone else who has an idea, for that matter. We don't trust it until we can see it and feel it."

At HP, it's a tradition that product-design engineers leave whatever they are working on out on top of their desk so that anyone can play with it. Walking around is the heart of their philosophy for all employees, and the trust level is so high that people feel free to tinker with the things their colleagues are inventing. One young engineer says: "You quickly learn that you ought to have something for people to play with. You are told probably on the first day that the fellow walking around playing with your gadget is likely to be a corporate executive, maybe even Hewlett or Packard." HP also talks about the "next bench syndrome." The idea is that you look around you to people working at the next bench and think of things that you might invent to make it easier for them to do their jobs.

Robert Adams, head of R&D at 3M, puts it this way: "Our approach is to make a little, sell a little, make a little more." McDonald's has more experimental menu items, store formats, and pricing plans than any of its competitors. In the course of our first three hours of interviewing at Dana, we heard mention of more than sixty different productivity experiments that were going on at one plant or another. P&G is, as we have said, especially well known for what one analyst calls its "testing fetish." Other examples from well-managed companies pour in daily. According to one analyst, "Bloomie's [Bloomingdale's] is the only large-volume retailer that experiments storewide." In fact, in response to that observation, an employee from Levi Strauss who was attending a recent seminar piped up and said, "You know that's where Levi got the faded jeans idea. Bloomie's was buying our jeans and bleaching them." Holiday Inns is said to have 200 test hotel sites in operation where they are continually experimenting with rooms, pricing, and restaurant menus. At the very successful Ore-Ida company, market tests, taste tests, pricing tests, and consumer panels are under way continuously, and the chief executive is as familiar with these tests and their results as he is with the financials.

The critical factor is an environment and a set of attitudes that encourage experimentation. This comment, by the man who invented the transistor, catches the quintessence of the experiment:

I lean more to being a believer of low cunning and expediency. . . . How do you go about starting a job? You have the people who read everything; they don't get anywhere. And the people who read nothing—they don't get anywhere either. The people who go around asking everybody, and the people who ask nobody. I say to my own people, "I don't know how to start a project. Why don't you step out and do an experiment?" You see, there is one principle here. You don't first start on something which is going to take six man-months before you get to the answer. You can always find something in which, in a few hours of effort, you will have made some little steps.

David Ogilvy likewise says there is no more important word than "test":

The most important word in the vocabulary of advertising is TEST. If you pretest your product with consumers, and pretest your advertising, you will do well in the marketplace. Twenty-four out of twenty-five new products never get out of test markets. Manufacturers who *don't* test-market their products incur the colossal cost (and disgrace) of having their products fail on a national scale, instead of dying inconspicuously and economically in test markets. Test your promise. Test your media. Test your headlines and your illustrations. Test your level of expenditure. Test your commercials. Never stop testing and your advertising will never stop improving. . . . Most young men in big corporations behave as if profit were not a function of time. When Jerry Lambert scored his first breakthrough with Listerine, he speeded up the whole process of marketing by dividing time into *months*. Instead of locking himself into *annual* plans, Lambert reviewed his advertising and his profits every month. The result was that he made $25,000,000 in eight years, where it takes most people twelve times as long. In Jerry Lambert's day, the Lambert Pharmacal Company lived by the month, instead of by the year. I commend that course to all advertisers.

Peter Peterson (now chairman of Lehman Brothers), speaking of the days when he was president of Bell & Howell, provides a lovely, concrete example of an experiment:

Have you heard of zoom lenses? One of the great advantages of being new in a company is that you are thoroughly unaware of what cannot be done. I thought a zoom camera was something that you used for football games. That was my image—an extraordinarily expensive object. One day I was in the lab, and there was a zoom lens. I had never seen one in my life, and I put it up to my eyes, and—well, it is a very dramatic thing. They explained to me that this was not applicable to consumer products, because it would cost a fair amount of money and so on. I asked, "What would it cost to make a camera for me—just one with a zoom lens on it?" They said, "Just one? Do you mean a crude modification? I think we would probably spend $500 on it." I said, "Well, suppose we do that; because my rates come pretty high, it will cost at least $500 for us to continue this discussion for another hour or two, so let's just do this." I took this camera home. At a dinner party that night, I put this zoom lens on the piano, and I asked everybody coming in if they wouldn't participate in a very sophisticated piece of market research; namely, to put the camera to their eye. To the man, the reaction was extraordinarily enthusiastic: "My, this is marvelous; I've never seen anything like this in my life." We did this for about $500. . . . If more industry would try out new ideas on a low-cost basis, perhaps their expectations of what the market will bear would go up.

Peterson's story contains several important messages about the experimenting mentality in business. The obvious one is the cost effectiveness of trying something as an alternative to analyzing everything. Less obvious is the ability of people to think more creatively—and at the same time concretely—with prototype in hand.

In his classic work *Language in Thought and Action,* S. I. Hayakawa captures the essence of the phenomenon when he points out that a cow is not a cow. Bessie the cow is not Janie the cow. He is talking about the importance of being able to leap from one level of abstraction to another—from cow to Bessie and Janie—in order to think clearly or communicate effectively.

For instance, one of us recently spent a pleasant weekend afternoon concocting homemade soap. The task is not too complex. The manual we used was clearly, even, at times, beautifully written. Yet we did a host of things wrong, we learned a score or more little tricks that will help next time around—all in just two or three hours. For example, exact alignment of temperatures between the lye mixture and the dissolved fats mixture is critical. The manual is clear on this, and provides lots of tips. But still we had problems; one pan was metal, shallow, and had a large surface area; the other container was glass, tall, and narrow. Shape and material differences, among other things, led to substantially different cooling rates at the critical moment. Only "feel" can help one to confront such complex phenomena quickly. The richness of the experience (in mathematical parlance, the number of variables surfaced and manipulated) that occurs *solely* when one is exposed tangibly to a subject, material, or process is unmatchable in the abstract, via paper analysis or description.

Thus, when "touch it," "taste it," "smell it" become the watchwords, the results are most often extraordinary. Equally extraordinary are the lengths to which people will go to avoid the test-it experience. Fred Hooven, protégé of Orville Wright, holder of thirty-eight major patents, and senior engineering faculty member at Dartmouth, describes a ludicrous, yet all-too-typical, case: "I can think of three instances in my career in which my client was making no progress on a complicated mechanical problem, and I insisted that the engineers and the technicians [model builders] be put in the same room. In each case the solution came rapidly. *One objection I remember being offered was that if we put the engineers in the same room with the shop it would get the drawings dirty.*" Hooven adds, in support of the overall point, "The engineer must have immediate and informal access to whatever facilities he needs to put his ideas into practice. . . . It costs more to make drawings of a piece than to make the piece, and the drawing is only one-way communication, so that when the engineer gets his piece back he has probably forgotten why he wanted it, and will find out that it doesn't work because he made a mistake in the drawings, or that it needs a small change in some respect, which too often takes another four months to make right."

So, via experimentation, it is much easier for people (e.g., designers, marketers, presidents, salesmen, customers) to think creatively about a product, or be creative about product uses, if a prototype, which is to say a low level of abstraction, is in hand. Thus, no amount of market research would have predicted the phenomenal success of the Apple II computer. We think it was the combination of a high-quality product and the emergence of an astonishing network of user groups, all playing with the machines and contributing new software almost daily, that made it such a success. No market research would have predicted that a woman we know would be the single biggest user of the Apple in her family; she, least of all, would have predicted that. It was starting her own business, based in her home, and having the Apple *there,* where she could try it and play with it at leisure, that made all the difference. Beforehand, had you told her about the wonders of word processing, she would have predicted (in fact, did predict) that she wouldn't use it. The concept was too abstract. Having the machine to play with, though, made her a convert.

That is why HP puts such emphasis on having its engineers leave their new experimental prototypes out where others can fool around with them. That is why Peterson's dinner-party market research on the zoom lens was, in fact, the most sophisticated marketing research imaginable.

Speed and Numbers

Alacrity and sheer numbers of experiments are critical ingredients to success through experimentation. Several years ago, we studied the successful versus the less successful wildcatters in the oil business. We concluded that if you had the best geologists, the latest in geophysical technique, the most sophisticated equipment, and so on, the success rate in wildcat drilling in established fields would amount to about 15 percent. Without all these pluses, the success ratio dips to around 13 percent. That finding suggests that the denominator—the number of tries—counts for a great deal. Indeed, an analysis of Amoco, recently revitalized to become the top U.S. domestic oil finder, suggests just one success factor: *Amoco simply drills more wells.* The company's head of production, George Galloway, says, "Most favorable results were unforeseen by us or anybody else. . . . That happens *if* you drill a lot of wells." We found the same phenomenon in minerals exploration. The critical difference between the unsuccessful exploration companies is a dramatic difference in the amount of diamond (bit) drilling that they do. Although diamond drilling *looks* expensive, it is the only way to find out what's really down there. The rest is all speculation, however well informed, by the geologists and geophysicists.

A former Cadbury's senior manager likewise underscores the value of speed and numbers. He recalls Cadbury's appointment of a new product development executive. The fellow looked at what was lying fallow in the development pipeline and blithely announced that there would be six new-product rollouts in the next twelve months. And six in the twelve months thereafter. Almost everything he planned to roll out had been in various states of limbo for two to seven years. He met his schedule, and three of the products are still big winners today. A veteran of the event commented, "You can cut the time to launch *at will,* if you just want to. He went through twelve in just twenty-four months. We wouldn't have done an iota better if we had taken five years to launch the same volume."

Peterson explains the rationale behind the Cadbury's phenomenon. An experiment, because it is a simple action, can be subjected to unreasonably tight deadlines. Under deadline pressure—*and* with manageable acts to perform—the impossible occurs regularly, it seems. Peterson comments:

It has been my observation that people often work on something for years and then some urgent situation comes up . . . and it suddenly comes through. Now, in one case we had an 8mm electric eye movie camera in development, and we anticipated it would take about three years to complete. Then one day the marketing vice president decided to try a different technique. He took something down to the engineers and said, "I just got an announcement that our competitors have an 8 millimeter electric eye camera!" Within 24 hours they had a completely different approach. I wonder just what is the role of urgency?

Speed means quick in (try it now) *and* also quick out. Storage Technology president Jesse Aweida's penchant for making decisions keeps the whole company in a state of constant experimentation. *Fortune* reports:

A disc drive . . . cost $1,500 more to make than it sold for. With characteristic dispatch, Aweida raised the price 50%, and when that didn't work, killed off the product, despite having invested $7 million in it. . . . He loathes inaction. As he told STC's national sales meeting last January, "I often believe that making a decision, even a bad decision, is better than making no decision at all." His ability to change course quickly has rescued the company from some of its bad decisions. Fortunately for STC, Aweida's vaulting ambition is balanced by his knack for quick correction.

There is a quality in experimentation as a corporate mind set that resembles nothing so much as a game of stud poker. With each card the stakes get higher, and with each card you know more, but you never really know enough until the last card has been played. The most important ability in the game is knowing when to fold.

With most projects or experiments, no matter how many milestones you set or PERT charts you draw, all you are really buying with the money invested is more information. You never know for sure until after the fact whether it has all been worthwhile or not. Moreover, as the project or experiment gets rolling, each major step becomes much more expensive than the last one—and harder to stop because of sunk costs and, especially, ego commitments. The crucial management decision is whether to fold. The best project management and experimenting management systems we have seen treat these activities more or less like poker. They break them up into manageable chunks; review quickly; and don't over-manage in the interim. Making it work simply means treating major projects as nothing more than experiments, which is indeed what all of them are, and having the poker player's mental toughness to fold one hand and immediately start another whenever the current hand stops looking promising.

Cheap Learning: Invisibility and Leaky Systems

Experimentation acts as a form of cheap learning for most of the excellent companies, usually proving less costly—and more useful—than sophisticated market research or careful staff planning.

Again, talking of his days at Bell & Howell, Peterson is quite clear
about this:

Before we let an idea get emasculated, and before we let any thoroughly rational appraisal of the idea convince us that it will not
work, we ask ourselves another question. Is there any way that we
can experiment with this idea at low cost? The experiment is the
most powerful tool for getting innovation into action and probably
is not as widely used as it should be in American Industry. . . . The
point I am trying to make is that if we can get the concept of the
experiment built into our thinking and thereby get evidence on a lot
of these "can'ts," "won'ts," "shouldn'ts," etc., more good ideas will
be translated into action. . . . Let me give you [an] example. Because we are not a large company we cannot afford to take massive
risks in spending millions of dollars promoting something without
knowing whether it will be effective or not. One day someone
walked in with an idea that, on the surface, was "preposterous."
Those who have read the Harvard marketing casebooks will know
every reason why this will not work: why not sell a $150 movie
camera [this was 1956] by direct mail? . . . Rather than say, "Gentlemen, this idea is preposterous," we tried to build in this notion:
"Let's examine some reasons it might work." Then we asked the
key question: "What would it cost us to try out the idea?" The cost
was only about $10,000. The point is that we could have spent
$100,000 worth of time over-intellectualizing this problem. . . .
Nine out of ten experts will tell you this idea just will not work. Yet
it did and is now a basis of an important and profitable new business for us. It is possible for us all to get a little pompous about the
power of an intellectual, rational approach to an idea that is often
extremely complex.

Another important property of the experiment is its relative in-
visibility. At GE, one term for experimenting is "bootlegging."
(The parallel term at 3M is "scrounging.") There the tradition of
squirreling away a little bit of money, a little bit of manpower, and
working outside the mainstream of the organization is time-honored. Huge GE successes, such as those mentioned earlier in engineered plastics and aircraft engines, have resulted directly from
bootlegging. The process has been essential to GE. In fact, a recent

analysis suggested that virtually every major GE breakthrough in the past couple of decades had its origins in some form of bootlegging. Several observers have said the same thing about IBM. One former colleague of the senior Mr. Watson goes so far as to suggest that a company's innovative health can best be measured by the amount of surreptitious bootlegging going on. Tait Elder, who headed 3M's New Business Ventures Division (NBVD), comments that planning, budgeting, and even control systems should be specifically designed to be "a little leaky." Lots of people need a way to scrounge money and play at the margin with budgets in order to pursue maverick programs.

Finally, and most important, is *the user connection*. The customer, especially the sophisticated customer, is a key participant in most successful experimenting processes. We will turn at length to this notion in the next chapter, but for the moment we will simply say that much of the excellent companies' experimentation occurs in conjunction with a lead user. Digital has more inexpensive experiments going on than any of its competitors. (HP and Wang are close on Digital's heels.) Each is *with* a user, *on* a user's premises.

The McDonald's experiments, obviously, are all done in conjunction with users—the customers. Many companies, on the other hand, wait until the perfect widget is designed and built before subjecting it—late in the game and often after millions of dollars have been spent—to customer scrutiny. The Digital, McDonald's, HP, 3M magic is to let the user see it, test it, and reshape it—very early.

The Experimenting Context

Just as we said that the ad hoc devices, such as task forces, won't work unless the environment supports fluidity and informality, experimenting won't work if the context is wrong. Management has to be tolerant of leaky systems; it has to accept mistakes, support bootlegging, roll with unexpected changes, and encourage champions. Isadore Barmash, in *For the Good of the Company*, presents a fascinating chain reaction whereby just one person, Sam Neaman, triggered an extravagantly successful experimenting process that, in the sixties, added millions of dollars to the bottom line for McCrory's stores. It is such a superb description of how a successful experimenting process gets going that we will quote Neaman—then executive without portfolio, later chief executive—at length:

I had no authority . . . but here was an opportunity. Here was a store that had lost so much money. I wanted to know what it took to make a good store. So I said to John [a store manager], "Look, we are going to bring into this store a group of people, a team, and you'll be the quarterback. You and they will go and visit all the competition in town and write up what you find. You'll check our merchandise and write it up. Every evening you'll hold classes with a blackboard and will have a consultation with everyone. . . . In addition, I'm gonna bring in the regional manager, merchandisers, buyers, and other store managers. I want to know the sum total of our know-how by taking a sampling of a group of people dedicated to finding out what they can do thinking together." For weeks they studied the store. They had a tough time agreeing with each other, but they did. The spirit was sky-high; excitement was beyond description. Why? For the first time they were given a chance to express themselves as individuals and as a group, each one giving the best that he knew. . . . Not a nickel was spent. Every change was made from what we had in the store. Floors were changed, aisles widened, walls painted. It was a new store, a pleasure to the eye.

What put that store across? They knew they had to visit all the competition and then look at our store with a cold eye. They applied what they learned. Up till then, they had to look at the eyeballs of the boss and guess what it was he wanted. All I did was ask them to use their senses and their heads, and I got a damn good store. Over the next two years, it reduced its losses and then started making money. After all the hustle-bustle, the whole company became aware of it. The chairman and his entourage came running to see what was happening. Now everybody jumped on the bandwagon. Now everybody wanted a district—every vice president, the executive vice president, even the chairman.

Show the people a way. That's what I did. I even had a place to send everyone. Indianapolis. "Go to Indianapolis in Indiana," I told them. "Go there, look at the store, and learn. It was put together by people like you, using spit and polish and only their own normal talents." A little while later, in the home office, I changed the pattern. To a variety chain vice president who was in charge of buying, I said, "All right, Joe, you don't have to go to the Midwest. Do me an Indianapolis right here in New York. You have seen what can be done. So go do an Indianapolis in Flushing. But I don't want you to copy it. We'll keep Indianapolis as a sort of school." I told him to give me his version of a good variety store in Flushing.

Well, several weeks later he invited me to the store and I found one of the most beautiful retail stores I have ever seen. I immediately invited a few others to see it. You never would have believed that his horrible store would be the attraction of the neighborhood and the jewel of the company. Sales began rising right away, and the store became our best in New York. But what it also did was to challenge the other home-office executives to go out and "Do an Indianapolis."

As the parent company began to brag more and more, I expanded the variations. I used the idea of the Indianapolis store as a visual aid. This meant devising a system of selecting one unit for improvement, getting the people to bring it into shape, then bringing others to see what they did so they could learn from it. This became a substitute for writing memos or giving instructions on the phone. Instead, I said, "Come look and see. This is the new company—nothing else is—this is it!" I instructed every district (10 to 15 stores) that it must have its own model store. Every district manager would have to reflect all his knowledge in one store and from that "Indianapolis" improve all the stores in his district. It would be his model, his manager's model, and the model for everyone who would look at it. The idea caught on like wildfire. They did it evenings, Sundays, holidays. The Sundays became big shindigs with beer and food provided by the store's restaurant manager. They had the year of their life getting the chain in shape, all 47 districts.

Neaman's description is more than just a story of lots of people experimenting; it's also a story of people allowed to stick out a little bit, people who start to feel like winners. Most important here, it's a story of the context that allows—indeed, actually encourages—

people to try things. Beyond what we've already talked about, there seem to be two important contextual aspects to the process of experimenting in companies.

The first is slightly forced but mostly natural diffusion, diffusion that builds on itself. The heart of the diffusion process is how one starts. "Beginnings are such delicate times," one sage commented. He is right. You start with the easy stuff, the things that are easy to change, and the places where your support base within the company is clear. We saw Neaman doing just that. Indianapolis was neither the biggest nor most visible store. But it was, under Neaman's tutelage, a store that was ripe to try something. A friend, Julian Fairfield, had as an early management job the problem of turning around a wire and cable plant that was performing miserably. "Everything was wrong," he said. "I didn't know where to start. So I started with housekeeping. It was the one thing everyone could agree on, and it was easy to fix. I figured if I became a fanatic on housekeeping, which was easy to improve, they would naturally begin to buy in to some other changes." They did.

Chase Manhattan Bank recently finished a major, successful adjustment of its retail (consumer) operation. The story was virtually the same. Management started with the regional manager who was most excited about doing something. Hers was not the biggest, the worst, nor the best region. It was simply one that was ripe for change. That regional manager tried things out, tested things, scored some visible wins. The saga diffused from one volunteer to the next. Only at the end did the most recalcitrant come on board. Similarly, McDonald's introduction of the breakfast menu started in the boondocks. A few franchisees picked it up and it then spread, over a two-year period, like wildfire. It now accounts for 35 to 40 percent of McDonald's revenues. At Bloomingdale's, the experimental process started much the same way: the easiest possible department to do over was the chairman's favorite, imported foods. That's where it started. Then came furniture. High fashion, which has gotten most of the subsequent attention but was the hardest to change, came last.

The process of building momentum by accumulating small successes is nicely described by consultant Robert Schaffer:

The essential idea is to focus immediately on tangible results—rather than programs, preparations and problem-solving—as the first step in launching performance improvement thrusts. . . . It is almost always possible to identify one or two specific short-term bottom line goals for which the ingredients for success are in place. . . . The *results-first* approach changes the whole psychology of performance improvement. . . . People must ask different kinds of questions. . . . Not, "What is standing in the way?" but rather, "What are some things we can accomplish in the next little while?" . . . Instead of trying to overcome resistance to what people are *not* ready to do, find out what they *are* ready to do. . . . Almost inevitably, when the managers successfully complete a project, they have many ideas about how to organize subsequent steps.

Schaffer describes, à la Neaman in Indianapolis, how to pick a manageable task. He suggests honing and honing until the doable emerges. "Select *one* branch whose manager seems *interested* in innovation and progress. Work with a team of sales people to increase sales on a *few* selected lines, perhaps in only some *selected* market sectors, by a *specific* percent in a matter of a *month* or six weeks. As they see tangible results, they are to . . . recommend how to expand the test" (italics ours).

Schaffer, like Neaman, Fairfield, Chase Manhattan, and Bloomingdale's, unearths a large bunch of variables. The experimenting process is almost revolutionary. It values action above planning, doing above thinking, the concrete above the abstract. It suggests, in a very Zen-like fashion, going with the flow: doable tasks, starting with the easiest and most ready targets, looking for malleable champions rather than recalcitrant naysayers. The image of a host of modest risk takers at Bloomingdale's, 3M, TI, Dana, McDonald's, GE, HP, or IBM comes to mind. The whole notion of risk taking is set on its ear. It becomes risky in the excellent companies not to take a little risk, not to "step out and do a little something." The management task becomes one of nurturing good tries, allowing modest failures, labeling experiments after the fact as successes, leading the cheers, and quietly guiding the diffusion process. The experiment is at the very heart of a new approach to managing, even in the midst of the most staggering complexity at a GE or IBM.

SIMPLIFYING SYSTEMS

Fluidity, chunking, and experimenting are interestingly abetted by the character of the excellent companies' formal systems. For instance, a junior colleague recently gave one of us a reading assignment in preparation for an interview with a client. He had put together a set of accumulated proposals that had come up to our client's division president. The shortest ran to fifty-seven pages. That's not the way it is at Procter & Gamble.

P&G systems are small in number and simple in construction, in harmony with the institution's no-nonsense approach to execution. Managers talk about "the grooves being deep and clear." Their systems are well oiled, well understood, to the point. At P&G the language of action—the language of the systems—is the fabled *one-page memorandum.*

We recently had breakfast with a P&G brand manager and asked if the one-page memorandum legend was really true. "It waxes and wanes,"* he said, "but I just submitted a set of recommendations to make a few changes to my brand's strategy. It ran a page and a quarter and got kicked back. It was too long." The tradition goes back to Richard Deupree, past president:

. . . Deupree strongly disliked any memorandum more than one typewritten page in length. He often would return a long memo with an injunction: "Boil it down to something I can grasp." If the memo involved a complex situation, he sometimes would add, "I don't understand complicated problems. I only understand simple ones." When an interviewer once queried him about this, he explained, "Part of my job is to train people to break down an in-

* For example, chairman Neil McElroy's historic memorandum of May 13, 1931, recommending brand versus brand competition, "bravely ran to three pages."

volved question into a series of simple matters. Then we can all act intelligently."

Ed Harness, P&G's recently retired chairman, echoes the tradition: "A brief written presentation that winnows fact from opinion is the basis for decision making around here."

The proliferation of MIS and forecasting models, the endless battles between numerous staffs—and the attendant "politicalization" of the problem-solving process—are among the reasons for growing unreliability. A one-page memo helps a lot. In the first place, there are simply fewer numbers to debate, and the ability to cross-check and validate twenty on one page, say, is easier than twenty times a hundred. It focuses the mind. Moreover, one stands on display. You can't reasonably hold someone responsible for getting a number wrong deep in Appendix 14. If there are only twenty numbers, on the other hand, accountability goes up automatically—and breeds reliability. Sloppiness is simply inconsistent with the one-page memo.

B. Charles Ames, past president of Reliance Electric and now president of Acme-Cleveland, makes a related point. "I can get a division manager to cough up a seventy-page proposal overnight," he says. "What I don't seem to be able to do is get a one-page analysis, a graph, say, that shows the trend and projection, and then says, 'Here are the three reasons it might be better; here are the three things that might make it worse.'"

John Steinbeck once said that the first step toward writing a novel is to write a one-page statement of purpose. If you can't get the one page clear, it isn't likely you'll get far with the novel. We are told that that is fairly conventional wisdom in the writing trade, but it apparently eludes most businessmen. It's little wonder that key assumptions get lost in a 100-page investment proposal. The logic probably is loose. The writing most likely is padded. The thinking is almost by definition shoddy. And, worse, the ensuing debate about the proposal among senior executives and reviewers is apt to be similarly unfocused.

A financial analyst once said of P&G, "They are so thorough it's boring." Another added, "They are a very deliberate, exacting company." Outsiders wonder how they can be all *that* thorough, deliberate, and exacting if reports are only a page long. Part of the answer lies in the struggle to get it all on that one page. Tradition has it that the typical first memo by an assistant brand manager or young brand manager requires at least fifteen drafts. Another part of the answer is that they have plenty of back-up analysis available, just like everyone else. The difference at P&G is that they don't inflict all those pages on one another. Still another compelling feature of the one-page cult is . . . less paper!

The power of the one-page memo is that its real impact goes much deeper than this partial list of traits. Apropos of curbing the paper chase and favoring action, Jorge Diaz Serrano, chairman of Pemex, the Mexican oil company, reports that he quit responding in writing to all written material and started using the phone; he aimed to establish a model of communication for the company. And Harry Gray, chairman of United Technologies, says: "I am known as a man who hates paper. When I first took over the job as chief executive, I called all the principal officers together in a room and told them of this insane dislike of paper. I have a phobia about it. I also told them that I had been burdened for one year in reading all of their carbon copies of what they considered to be important correspondence. I directed them to cease and desist and not to send me another piece of paper except for one-page memos."

Charles Ames, talking about his earlier experience at Reliance, speaks of the love affair with complex systems that often hides an inability to manage the basics: "We had planning systems of every sort from very long-term strategic systems to short-term ones. But we couldn't predict what we were going to sell next *month*. I dismantled the five-year planning system, and went to a one-year planning system, and next to a quarterly system. We ended up running the company on a thirty-day system for a year or so. Only then did we learn to get the numbers right. Eventually we built back up to a long-term system, though never back to the epic proportions of the one we'd had originally."

Contrary to Ames's initial experience, Emerson Electric, Dana,

TI, and other companies foster quick response through focus on *one or two closely watched numbers.* For example, a *New York Times* report on Emerson Electric notes: "Division presidents and their top lieutenants are put under the microscope at headquarters every month by their group vice president. The focus is more on the present than the future. Three items—inventories, profits, and sales—form a crucible for managers. They are told that what they've got to do is make sure the profit is delivered each month, each quarter, and—ultimately—the full year." Similarly, a *Management Today* article on Dana states: "Although head office does not require much in the way of written reports, it does need a certain minimum of information. The most important item is the revenue figure. In the old days it used to come up, along with much else, in an actual-against-budget tabulation by the 20th of the following month. Under the current system, the divisions transmit to head office, by phone or telex, their invoice total, and approximate profit earned, at the end of each working day."

Virtually any system can be cleaned up and made simple. Some watchwords at TI are: "More than two objectives is no objectives," and, "We got over the scoring phase in the early seventies." Yes, TI is a systems-driven company; ex-chairman Haggerty spent a decade instilling what he calls the "language" of the Objectives, Strategies, and Tactics system. But the principal OST thrust is on fostering informal communications and personal accountability—and there's no better window on TI's techniques than the seemingly mundane two-objectives point. Most MBO systems we've run across include up to thirty annual objectives for a single manager. It's obvious that no one gets more than a handful of activities done every few months. TI simply recognizes this fact: "We've been through it all. Each manager used to have a bunch of objectives. But gradually we trimmed and trimmed and trimmed. Now each PCC manager [Product Customer Center, the TI equivalent of a division] has *one* milestone a quarter. That's it. You *can*—and we do—expect someone to get one thing done."

Others have instituted similar routines. Chairman John Hanley of Monsanto (P&G-trained, interestingly) says: "Three to five objectives [a year] is a maximum." HP's John Young echoes Hanley: "In our strategic reviews, the critical point is the division general manager's three to five objectives [for the year]. We really don't need the financials. The only reason that I use them is to keep the division managers happy. If they get those objectives right, the financials will follow." The nature of the HP objectives is important to action too—and, again, so different from those in the nonexcellent companies. Objectives at HP are *activities,* not abstract financials over which the manager has little control; for instance: "Get the plant in Eugene, Oregon, up to 75 percent capacity by March 15, " or, "Get the sales force in the Western Region spending 50 percent of their time calling on customers of type X rather than type Y by October 31."

While one-page memos, honest numbers, and focused objectives are the systems traits of the excellent companies, the context is equally important. The trouble is that the context can be observed only as the sum of scores of seemingly mundane traits. Plenty of companies have tried all the traits and the systems—brief communications, fact-based decision making, management by objectives. But they try, don't succeed initially, and then give up; another gimmick down the drain. Few persist with systems design until they've gotten the trade-off between simplicity and complexity right. P&G has been deepening the grooves of its one-page communications system for forty years.

THE ACTION ORIENTATION

There is no more important trait among the excellent companies than an action orientation. It seems almost trivial: experiments, ad hoc task forces, small groups, temporary structures. Whether it's the introduction of IBM's System 360 (a seminal event in American business history) or a three-day ad hoc task force at Digital, these companies, despite their vast size, are seldom stymied by overcomplexity. They don't give in and create permanent committees or task forces that last for years. They don't indulge in long reports. Nor do they install formal matrixes. They live in accord with the basic human limitations we described earlier: people can only handle a little bit of information at one time, and they thrive if they perceive themselves as even somewhat autonomous (e.g., experimenting modestly).

The major complaint about organizations is that they have become more complex than is necessary. Refreshingly, the excellent companies are responding by saying: If you've got a major problem, bring the right people together and expect them to solve it. The "right people" very often means senior people who "don't have the time." But they do, somehow, have the time at Digital, TI, HP, 3M, IBM, Dana, Fluor, Emerson, Bechtel, McDonald's, Citibank, Boeing, Delta, *et al.* They have the time in those institutions because those companies aren't transfixed with organization charts or job descriptions or that authority exactly matches responsibility. Ready. Fire. Aim. Learn from your tries. That's enough.

6

Close to the Customer

Probably the most important management fundamental that is being ignored today is staying close to the customer to satisfy his needs and anticipate his wants. In too many companies, the customer has become a bloody nuisance whose unpredictable behavior damages carefully made strategic plans, whose activities mess up computer operations, and who stubbornly insists that purchased products should work.
—Lew Young, Editor-in-Chief, *Business Week*

That a business ought to be close to its customers seems a benign enough message. So the question arises, why does a chapter like this need to be written at all? The answer is that, despite all the lip service given to the market orientation these days, Lew Young and others are right: the customer is either ignored or considered a bloody nuisance.

The good news from the excellent companies is the extent to which, and the intensity with which, the customers intrude into every nook and cranny of the business—sales, manufacturing, research, accounting. A simple message permeates the atmosphere. All business success rests on something labeled a sale, which at least momentarily weds company and customer. A simple summary of what our research uncovered on the customer attribute is this: the excellent companies *really are* close to their customers. That's it. Other companies talk about it; the excellent companies do it.

No existing management theory helps much in explaining the

role of the customer in the prototypical excellent company. At most, recent theory talks about the importance of the external environment in influencing the institution. It misses by a mile, however, the intensity of customer orientation that exists within the top performers, and that intensity seems to be one of the best kept secrets in American business.

The case was nicely expressed by HP's John Doyle (head of R&D). We were discussing sustaining business values. He said that the only posture that has a chance of surviving the ravages of time is one that is unfailingly externally focused: "The only way you're going to survive in the long haul is if everybody's out there scratching, looking for things to do to get the next product generation into the customer's premises."

In observing the excellent companies, and specifically the way they interact with customers, what we found most striking was the consistent presence of *obsession*. This characteristically occurred as a seemingly unjustifiable overcommitment to some form of quality, reliability, or service. Being customer-oriented doesn't mean that our excellent companies are slouches when it comes to technological or cost performance. But they do seem to us more driven by their direct orientation to their customers than by technology or by a desire to be the low-cost producer. Take IBM, for example. It is hardly far behind the times, but most observers will agree that it hasn't been a technology leader for decades. Its dominance rests on its commitment to service.

Service, quality, reliability are strategies aimed at loyalty and long-term revenue stream growth (and maintenance). *The point of this chapter, and a wonderful concomitant to a customer orientation, is that the winners seem to focus especially on the revenue-generation side*. The one follows the other.

SERVICE OBSESSION

Although he's not a company, our favorite illustration of closeness to the customer is car salesman Joe Girard. He sold more new cars and trucks, each year, for eleven years running, than any other hu-

man being. In fact, in a typical year, Joe sold more than twice as
many units as whoever was in second place. In explaining his secret
of success, Joe said: "I send out over thirteen thousand cards every
month."

Why start with Joe? Because his magic is the magic of IBM and
many of the rest of the excellent companies. It is simply service,
overpowering service, especially after-sales service. Joe noted,
"There's one thing that I do that a lot of salesmen don't, and that's
believe the sale really begins *after* the sale—not before.... The
customer ain't out the door, and my son has made up a thank-you
note." Joe would intercede personally, a year later, with the service
manager on behalf of his customer. Meanwhile he would keep the
communications flowing:

Joe's customers won't forget him once they buy a car from him; he
won't let them! Every month throughout the year they get a letter
from him. It arrives in a plain envelope, always a different size or
color. "It doesn't look like that junk mail which is thrown out before it is even opened," Joe confides. And they open it up and the
front of it reads, 'I LIKE YOU.' Inside it says 'Happy New Year from
Joe Girard.'" He sends a card in February wishing the customers a
"Happy George Washington's Birthday." In March it's "Happy St.
Patrick's Day." They love the cards. Joe boasts, "You should hear
the comments I get on them."

Out of context, Joe's 13,000 cards sounds like just another sales
gimmick. But like the top companies, Joe seems genuinely *to care*.
Said Joe: "The great restaurants in the country have love and care
coming out of their kitchens ... and when I sell a car, my customer's gonna leave with the same feeling that he'll get when he walks
out of a great restaurant." Joe's sense of caring continued to shine
through after the sale: "When [the customer] comes back for service, I fight for him all the way to get him the best.... You've got
to be like a doctor. Something's wrong with his car, so feel hurt for
him." Moreover, Joe has cared about every customer as an individual. He doesn't think statistically, but emphasizes that he has sold
"one at a time, face-to-face, belly-to-belly." "They are not," he
said, "an interruption or pain in the neck. They are my bread and

butter." We introduce this section with Joe because he has acted, as well as anyone, as if the customer really does count.

"I was at a meeting of sales managers with Mr. Watson [Senior] one time," says Gordon Smith, recently retired from Memorex. "The purpose was to assess some customer problems. On the front table there were eight or ten piles of papers, identifying the source of problems: 'manufacturing problems,' 'engineering problems,' and the like. After much discussion, Mr. Watson, a big man, walked slowly to the front of the room and, with a flash of his hand, swept the table clean and sent papers flying all over the room. He said, 'There aren't any categories of problems here. There's just one problem. Some of us aren't paying enough attention to our customers.' He turned crisply on his heel and walked out, leaving twenty fellows wondering whether or not they still had jobs."

In *A Business and Its Beliefs,* Thomas J. Watson, Jr., talks about the ideas that helped build the company. He makes this cogent point about service:

In time, good service became almost a reflex in IBM. . . . Years ago we ran an ad that said simply and in bold type, "IBM Means Service." I have often thought it was our very best ad. It stated clearly just exactly what we stand for. *We want to give the best customer service of any company in the world* . . . IBM's contracts have always offered, not *machines* for rent, but machine *services,* that is the equipment itself and the continuing advice and counsel of IBM's staff.

Like Joe Girard, IBM is fanatic about its service beliefs. In most companies "assistant to" functionaries are usually bag carriers, paper shufflers, gofers. Not so at IBM. There, some of the best salesmen are made assistants to the company's top officers. While people are in this position, they spend their entire, typical three-year stint doing only one thing—*answering every customer complaint within twenty-four hours.* (In the field, the swarm effort is equally notable. A Lanier data-processing executive in Atlanta, a competitor in some areas, swears by IBM mainframes: "I remember the last time we had trouble. In hours the horde descended, from everywhere. They called in about eight experts on my problem. At least four were from Europe, one came from Canada, one from Latin America. That's just where they happened to be.")

The eerie part of the IBM story on service is the absence of chinks in the armor. Recently, in a one-week period, one of us (1) sat next to a twenty-five-year-old Oakland-based IBM salesperson on a flight from New York to San Francisco, (2) talked to a senior AT&T executive with an IBM background, (3) talked to a Memorex executive who had been an IBM manufacturing executive, (4) discussed an IBM sales decision with a hospital administrator, and (5) talked with a young ex-IBM salesman in a classroom setting. They didn't look alike; they ranged from an attractive young black woman to a grizzled fifty-year-old. But they did talk alike. All these people agreed that IBM has had problems—software, even quality sometimes. But all also agreed, using practically the same words, that IBM's service and reliability are unmatched. What's so impressive is the depth and consistency of their belief that IBM really *cares* about service.

Reinforcing examples abound. Our office is on the forty-eighth floor of the Bank of America World Headquarters; consequently, we come into contact with many B of A executives. One friend had been put in charge of operations for the World Banking Division. He told us that when he started the job, about three months prior to our discussion, he really had but one principal objective, to wean the bank from total dependency on IBM. "Get some stuff from Amdahl, for instance." He continues, "I had been on the job, I'd guess, about four weeks, when I walked in one morning and there was a huge proposal on my desk called 'Systems Requirements for the Eighties.' I looked at it. It was from my IBM account executive. I didn't want it. I called him up and asked, 'Why the hell are you doing this to me?' He was very direct and to the point. He said, 'That's the way we control the customer!'"

When you listen to corporate marketing vice president Buck Rodgers speak, as we did recently, there's a feeling of *déjà vu,* and you suddenly realize you are listening to the modern incarnation of Watson insisting on the Golden Rule (of service). Rodgers states that every proposal to a customer should be "overwhelmingly cost-justifiable from the customer's standpoint." (An ex-IBMer we know laments, "An IBM salesman always sells the cheapest product that will get the job done," adding that he wishes the same could be said of his present company. "I can't believe it," he says of the latter. "They try to sell them the Brooklyn Bridge. They act like there's no tomorrow.") Rodgers comments that IBM is "customer- and market-driven, not technology-driven." He says he wants salesmen to "act as if they were on the customer's payroll," and he talks of putting "*all IBM* resources at the customer's disposal." Finally, he notes that "getting the order is the *easiest* step; after-sales service is what counts." He adds that IBM keeps its sales branches small (maximum 100 people) so "we can be easy to do business with." He notes, conclusively, that "we must be constantly in touch."

To make sure it is in touch, IBM measures internal and external customer satisfaction on a monthly basis. These measures account for a large share of incentive compensation, especially for senior management. Employee attitude surveys are taken every ninety days, and a check is kept on employee perceptions of the way customer service is being maintained.

The corporate officers at IBM still make sales calls with great regularity. In New York once of us recently ran into a senior financial officer who makes customer calls and insists that all his people do so as well: "How's someone going to design a receivables policy if he doesn't know the customer?" Chairman John Opel underscores the point: "You have to remember who pays the bills. No matter what the primary discipline—finance, manufacturing—you have to know and experience the excitement of sales. That's where you really see things happen."

IBM backs its close-to-the-customer beliefs with intensive training. Basic sales training is fifteen months: 70 percent of the time is spent in the branch, and 30 percent in university-like settings. Advanced training follows like clockwork. For example, more than 1,000 people per year go through the President's Class. It is conducted by eight Harvard professors and six IBM professors, and its purpose is to "teach people how customer presidents think." Roughly another 1,000 salesmen go through a financial officer's course, also run jointly with Harvard. They learn how financial officers think. It is part of a program that adds up to an estimated 15 days spent on formal training for everyone, every year, regardless of seniority.

There is a tough side to IBM's emphasis on service. Account representatives have "full liability" for the equipment in place. For example, suppose you are an account representative and you call on an account tomorrow morning and are told, at your first meeting, that some of the IBM equipment recently in place has to be taken out. Even though your predecessor had been the sales representative for the last ten years (and is therefore the likely cause of the withdrawal), Rodgers adds that you would still be docked, out of bonus *and* salary, for the full amount of commission paid to the previous sales representative for placing the original order. Needless to say, this system reflects the depth of IBM's commitment to after-sale service and the importance of continuing customer relationships. Rodgers emphasizes, "It keeps the person involved with today's customer from the customer satisfaction side." Jacques Maison-Rouge, head of IBM World Trade, underscores the point: "IBM always acts as if it were on the verge of losing *every* customer."

Other tough systems include "joint loss reviews." Regional and branch people are brought in monthly to discuss account losses. In addition, the president, chairman, and senior officers all receive *daily* reports of lost accounts. Notes a senior ex-IBMer, "It's astonishing. I remember losing a big account once. I hadn't even gotten back to the office from the meeting when the phone was ringing off the hook. 'What happened? Let's talk about it.' It seemed like half of corporate descended on me the next day. To this day I have no idea how they found out so fast." Ex-IBMers are astonished by the absence of such vigorous systems in their new companies. One who is now an executive vice president at a competitor commented recently with dismay, "I can't believe it. The chairman doesn't even keep a *list* of our own top one-hundred customers."

Nevertheless, if you look hard enough, you can almost always find someone who does it better; for example, in a few market niches, *Lanier* outservices even IBM. A friend who heads up the word-processing business in a major corporation was talking about the slowness of diffusion of the office-of-the-future concept. He said that a problem is that everyone calls one of the principal components, the so-called smart typewriter, a "word processor." He says, "There sure as heck isn't a bigger put off to the user, the secretary, or any bigger threat than that term." Is there anybody who doesn't call it a word processor? Just one we know of: Lanier. And the last time we looked, little Lanier had beaten giant competitors IBM, Xerox, Wang, and about a hundred others in stand-alone word processors. They were share leaders and had solid margins to boot. They call their machines the "No Problem Typewriter." The label hints at Lanier's customer orientation. Lanier lives, sleeps, eats, and breathes customers. In fact, a colleague commented that being around Lanier executives is like being in a football locker room at half-time in a close game. The high-volume chatter focuses ceaselessly on sales, customers, and head-on-head competition with competitors.

It starts, as at IBM, with role models. Wesley Cantrell, the president of Lanier, exudes customer orientation. The top Lanier executives all make sales calls once a month. Lanier's customer orientation also stresses product simplicity and "friendliness." Cantrell has been heavily influenced by his early stint as a salesman. He sold 3M office copy products. He says that Kodak's instruction booklet was fifteen pages long, whereas 3M's instructions took up only one sheet of paper. "Their instruction manual was my best sales tool," he comments.

Lanier wants to make its product easy on the user, and it works. A recent Harvard Business School doctoral dissertation contrasted Xerox, Wang Labs, and Lanier in a study of adaptation. It found that Lanier's orientation was the nearest to the ultimate user, the secretary. The result was very rapid adoption of features that were attractive to secretaries.

With an excessively short service turnaround and a short on-site service time, Lanier beats even IBM at IBM's service game. Both service turnaround and time on site are measured incessantly by Lanier management. To accomplish quick service, they spend money. They "overequip" their service people. Investment in the tools and testing devices that the Lanier service representative carries around is substantially above the industry average. Lanier also tries to outdo IBM on response to complaints. The company contends that it answers all complaints within four hours, and the president handles a big share himself. (He adds, "And I charge my regional sales and service people at *my* hourly rate for handling the problem.") He likes to beat the four-hour standard, and he says: "Of course, the No-problem Typewriter makes that easier to do."

Perhaps our favorite example of service overkill is *Frito-Lay*. We have been exposed to a good deal of micro-economic theory, and it sometimes appears that there is only one thing that economists are absolutely sure of after several hundred years of labor: wheat farmers in perfectly competitive markets don't have high margins. We don't have any excellent wheat farmers in our survey, but we got pretty close. Potato chips and pretzels ought to be the classic undifferentiated commodity. Like wheat farmers, potato chip manufacturers ought not to have high margins or shares. But Frito-Lay, a subsidiary of PepsiCo, sells well over $2 billion worth of potato chips and pretzels every year, owns market shares that run into the 60s and 70s in most of the country, and has margins that are the envy of the food industry. Why?

What is striking about Frito is not its brand-management system, which is solid, nor its advertising program, which is well done. What is striking is Frito's nearly 10,000-person sales force and its "99.5 percent service level." In practical terms, what does this mean? It means that Frito will do some things that in the short run clearly are uneconomic. It will spend several hundred dollars sending a truck to restock a store with a couple of $30 cartons of potato chips. You don't make money that way, it would seem. But the institution is filled with tales of salesmen braving extraordinary weather to deliver a box of potato chips or to help a store clean up after a hurricane or an accident. Letters about such acts pour into the Dallas headquarters. There are magic and symbolism about the service call that cannot be quantified. As we said earlier, it is a cost analyst's dream target. You can always make a case for saving money by cutting back a percentage point or two. But Frito management, looking at market shares and margins, won't tamper with the zeal of the sales force.

Frito simply lives for its sales force. The system succeeds because it supports the route salesman, believes in him, and makes him feel essential to its success. There are about 25,000 employees in the company. Those who are not selling live by the simple dictum, "Service to Sales." While the plant manager, to pick an example, is clearly evaluated on the traditional basis of whether or not he makes his cost budget, when the sales force is in a crunch he won't hesitate to run the plant overtime to make sure sales gets what it needs. If he doesn't, he'll hear about it from all quarters, like our IBM friend who lost his big account.

The best outside analysis of the close-to-the-customer-through-service concept that we have come across is a 1980 effort performed by Dinah Nemeroff of Citibank. She had eighteen respondents, including American Airlines, Disney Productions, McDonald's, Westin, Hertz, and IBM. One of Nemeroff's most interesting findings is that people in these different but service-intensive companies use the same language in describing themselves. She notes, "They discuss service issues in identical words."

Nemeroff finds three principal themes in an effective service orientation: (1) intensive, active involvement on the part of senior management; (2) a remarkable people orientation; and (3) a high intensity of measurement and feedback. As we have found over and over, it starts with senior management. Nemeroff neatly calls it "service statesmanship." Senior executives exercise that statesmanship through personal example. Their commitment starts with a company philosophy. In fact, many of the companies in her survey devoted an *explicit* part of their mission statement to discussion of service. And in many of these companies, service excellence was viewed as the prime objective. With service as their top goal, they said that "profitability naturally follows," which reinforces the revenue-generation point made at the beginning of this chapter.

Nemeroff picked up numerous examples of management style traits that reinforced the service philosophy. She found that top managers treated the service problems as "real time" issues—issues that deserve their immediate personal attention. She found that top management directly intervenes, ignoring the chain of command, in decisions about service. These mangers have frequent regular meetings with junior professionals who respond to customer mail. They pen "marginal notes on customer correspondence," and "engage in dramatic service delivery gestures to increase visibility to customers." (And, we would add, to reinforce this service message throughout their own organizations.)

Of another aspect of top management style, Nemeroff makes a crucial and surprisingly subtle point: "Interviewed executives believe they must maintain a long-term view of service as a revenue builder." This point is all too often missed in big American companies. Profit objectives, while very necessary, are internally focused and certainly do not inspire people by the thousands way down the line. Service objectives, on the other hand, are almost without fail meaningful to down-the-line employees. A strong sense of personal accountability among down-the-line employees is crucial. And one knows that has been accomplished when someone in the field says, as did one of Nemeroff's respondents, "Each one of us *is* the company."

Nemeroff makes the important connection that "customer relations simply mirror employee relations." Inseparable from the way these service-oriented companies manage their people is the intensity of measurement and feedback systems. Perhaps her most significant finding in this regard was that new rewards and incentive programs are in continuous preparation. For example, one respondent in her survey said that "service-incentive programs are changed at least every year to keep them fresh, and most are homegrown by local management." This really struck us in all aspects of the work of the excellent companies. Programs for people—incentive programs, training programs, or simple hoopla—undergo continuous retuning, much as product development does. No practice is expected to have impact forever, and programs for people have life cycles just as products do, maybe even shorter ones.

One of the best examples of service through people is *Walt Disney Productions*. In fact, many rate Disney and McDonald's as the two best mass service providers in America—or the world. Red Pope, a long-time Disney observer and writer, comments: "How Disney looks upon people, internally and externally, handles them, communicates with them, rewards them, is in my view the basic foundation upon which its five decades of success stand . . . I have come to observe closely and with reverence the theory and practice of selling satisfaction and serving millions of people on a daily basis successfully. It is what Disney does best."

Pope's observations on Disney are a clear validation of the Nemeroff study. For example, intense management involvement is highlighted at Disney by an annual week-long program called "cross utilization." According to Pope, this program entails Disney executives' leaving their desks and their usual business garb. They don a theme costume and head for the action. "For a full week, the boss sells tickets or popcorn, dishes ice cream or hot dogs, loads and unloads rides, parks cars, drives the monorail or the trains, and takes on any of the 100 on-stage jobs that make the entertainment parks come alive."

The service-through-people theme at Disney starts, as it does in many of the excellent companies, with a special language. There is no such thing as a worker at Disney. The employees out front are "cast members" and the personnel department is "casting." Whenever you are working with the public, you are "on stage." For example, two of Red Pope's children, aged sixteen and eighteen, were hired by Disney World in Orlando to take tickets. For this seemingly mundane job, four eight-hour days of instruction were required before they were allowed to go on stage. They learned about Guests—not lower-case "c" customers, but upper-case "G" Guests. Pope asked his children why it had taken four days to learn how to take tickets, to which they replied: "What happens if someone wants to know where the restrooms are, when the parade starts, what bus to take to get back to the campgrounds? . . . We need to know the answers and where to get the answers quickly. After all,

Dad, we're on stage and help produce the Show for our Guests. Our job every minute is to help Guests enjoy the party."

People are brought into the culture early. Everyone has to attend Disney University and pass "Traditions I" before going on to specialized training. Pope says:

Traditions I is an all-day experience where the new hire gets a constant offering of Disney philosophy and operating methodology. No one is exempt from the course, from VP to entry-level part-timers. . . . Disney expects the new CM [cast member] to know something about the company, its history and success, its management style before he actually goes to work. Every person is shown how each division relates to other divisions—Operations, Resorts, Food and Beverage, Marketing, Finance, Merchandising, Entertainment, etc. and how each division "relates to the show." In other words, "Here's how all of us work together to make things happen. Here's your part in the big picture."

The systems support for people on stage is also dramatic. For example, there are hundreds of phones hidden in the bushes, hot lines to a central question-answering service. And the amount of effort put into the daily clean-up amazes even the most calloused outside observers. In these and scores of other ways, overkill marks every aspect of Disney's approach to its customers.

Whether or not they are as fanatic in their service obsession as Frito, IBM, or Disney, the excellent companies all seem to have very powerful service themes that pervade the institutions. In fact, one of our most significant conclusions about the excellent companies is that, *whether their basic business is metal bending, high technology, or hamburgers, they have all defined themselves as service businesses.*

AT&T executive vice president Archie McGill is an ex-IBM executive. He goes farther and makes a nice distinction between broad-based service standards and what he calls a "customer focus" (a true service focus). The latter, he says, means "acknowledging that *every* individual perceives service in his own terms." Overmeasurement of service (e.g., scores of variables) may actually detract from it, he adds. One loses sight of the individual customer.

Suppose you have a "ninety-five percent standard." McGill asks, "What about the five percent? Even though one hundred percent may be theoretically unattainable, the business *ought to act as if any failure is intolerable.*"

Boeing is another excellent example. To be sure, the company manufactures airplanes, but what makes it outstanding is its service orientation. Says a *Wall Street Journal* analyst of Boeing:

Nearly every operator of Boeing aircraft has a story about the company's coming through in a pinch. When tiny Alaska Airlines needed landing gear that could put a jet down on a dirt strip, Boeing was there. When Air Canada had a problem with ice clogging in some air vents, Boeing flew its engineers to Vancouver, where they worked around the clock to solve the problem and minimize disruption of the airline's schedule. Boeing's attention to customer relations has paid off. In December 1978, Alitalia lost a DC9 airliner into the Mediterranean Sea and the Italian national carrier vitally needed a replacement aircraft. Umberto Nordio, Alitalia's president, telephoned T. A. Wilson, Boeing's chairman, with a special request: could Alitalia quickly get delivery on a Boeing 727? At the time there was a two-year wait for such aircraft, but Boeing juggled its delivery schedule and Alitalia got the plane in a month. Mr. Nordio returned the favor six months later, when Alitalia cancelled plans to buy McDonald Douglass DC10s and ordered nine 747 Jumbos [from Boeing], valued at about $575 million.

In talking about its amazing metamorphosis from a company that was primarily dependent on the military to a company that is primarily commercial, Boeing says of itself in the book *Vision,* "We've tried to build a team that is customer oriented. We came to a realization that, if we were going to succeed in commercial business, the important ingredient was the customer. We can't let the airline say—as it sometimes has—'the only time you are interested in our problem is when you're trying to sell us a new airplane.' It has taken us a long time to recognize the customer's problems. Now [this] point of view is beginning to percolate through the whole organization."

We can't conclude this discussion without briefly mentioning one issue that's paramount to many observers: can you spend too much on service? Of course, in an absolute sense, one can spend too much. But if yes is the answer absolutely, then we would say no directionally. That is, just as there are "too many" champions at 3M and "too many" divisions at HP or J&J, according to rational analysis, *almost every one of our service-oriented institutions does "overspend" on service, quality, and reliability.* As David Ogilvy reminds us: "In the best institutions, promises are kept no matter what the cost in agony and overtime." It holds for advertising, for computers, for typewriters, for amusement rides, and for pretzels.

Finally, we observed that the customer orientation is an intense motivator. We recently ran into a former J&J accounting staff employee who is now a senior vice president at Chase Manhattan Bank. He recalled: "Within the first couple of weeks, I made sales calls. It's typical. J&J says, in effect, if you can't understand the customers, you won't understand the business." Another friend tells a mundane but similar story:

I was at the Pentagon in the Office of the Chief of Naval Operations. I had a bunch of GS-11 and -12 civil servants [middle managers] working for me on some parts of the O&M [Operations and Maintenance] budget. I was always distressed that they were so demotivated towards work, but so animated in general. A lot of them were selling real estate or running other small businesses on the side. I had one "expert," though, who really was turned on. I only later realized the key. Because of his skill at shifting resources and finding extra pots of money, I'd often send him on two- or three-day temporary assignments down to Norfolk. He'd work with Fleet people and figure out some way to get them enough fuel for some extra maneuvers or whatever. I now realize he was simply the only one who had real "customer contact." He saw ships and the people who drove them. The numbers weren't abstractions to him. His actions had measurable, better yet tangible, effects. In retrospect there are a hundred things I could have done to have made that experience commonplace for all my people.

In our experience with the better-run companies, there is no part of an enterprise that can't be touched by the customer. Caterpillar sends people from the plants out to the proving grounds to watch the big machines at play. Citibank lets "back room" operations people regularly visit customers and account officers to solve operational problems directly. 3M insists that its most basic R&D people regularly visit customers; so does HP. In these ways, the service orientation becomes tangible for all hands. "Each of us *is* the company," comes to take on real meaning.

QUALITY OBSESSION

We've mentioned that many of our excellent companies are obsessed by service. At least as many act the same way over quality and reliability. A superb example is *Caterpillar Tractor*. Caterpillar offers customers forty-eight-hour guaranteed parts delivery service anywhere in the world; if it can't fulfill that promise, the customer gets the part free. That's how sure Cat is, in the first place, that its machines work. Once again, we are looking at a degree of overachievement that in narrow economic terms would be viewed as a mild form of lunacy; lunacy, that is, until you look at Caterpillar's financial results.

An article in *Fortune* states simply: "The company's operating principles seem to be an individual version of the Boy Scout law: the main principles are excellence of quality, reliability of performance, and loyalty in dealer relationships. Caterpillar has zealously pursued the goal of building a better, more efficient crawler tractor than anybody else in the world." A *Business Week* analyst concurs: "Product quality is something Cat people hold as close as a catechism." When we say anything about Cat in the presence of two senior agriculture executives we know, they both become almost misty-eyed with reverence. Likewise, one of us remembers Cat from days of ordering construction equipment for the Navy in Vietnam. We would go to almost any ends, stretching the procurement regulations to the limit, to specify the always more expensive Cat equipment. We had to, for we knew our field commanders would string us up if we didn't find a way to get them Cat. When you're airlifting bulldozers into unfriendly territory for the purpose of building short airstrips behind enemy lines, you want machinery that works—all the time.

In the case of Caterpillar, close to the customer also means close to the dealer. Said former president and chairman William Blackie, "We have a tremendous regard for our dealers. We will not bypass or undercut them. Some of our competitors do and their dealers quit. Caterpillar dealers don't quit; they die rich." Beyond dollars, Cat dealers are treated like "members of the family." For example, *Business Week* reports: "The Company even conducts a course in Peoria to encourage dealers' children to remain in the business. Executive vice-president for marketing E. C. Chapman recalls, 'We had a dealer's son who was studying for the ministry and had a secondary interest in music. By the time we sent him home, he changed his career plan. He has become one of our most successful dealers.'"

William Naumann, ex-chairman of Caterpillar, says that from the very beginning of the expansion of Caterpillar's business just after World War II, a fundamental decision was made that has had lasting impact on the way it conducted all its business. "We adopted a firm policy that a Caterpillar product or component—no matter where it was built—would be the equal in quality or performance of the same product or component built at any other location, whether in this country or abroad." He states that "users can count on the availability of replacement parts regardless of where they operate—an important consideration in a highly mobile industry. We have no orphans."

Naumann believes that this decision on reliability, quality, and uniformity has been an enormous unifying force in the company's development. "A machine made at one plant is the counterpart of the same machine made at any other plant, and parts are interchangeable throughout the world."

Another company that overkills on the quality dimension is *McDonald's*. For years, its theme has been "Quality, Service, Cleanliness, and Value": Q.S.C.&V. Says founder Ray Kroc, "If I had a brick for every time I've repeated the phrase Q.S.C.&V. (Quality, Service, Cleanliness, and Value), I think I'd probably be able to

bridge the Atlantic Ocean with them." Since the early days of the
organization, all of its stores have been regularly measured on their
performance in these categories, and the Q.S.C.&V. measure determines a big chunk of the store manager's compensation. Consistent
failure to meet McDonald's high Q.S.C.&V. standards can get
store managers fired or cause loss of a franchise.

Ray Kroc and other members of the top management team are
legendary for personally inspecting stores on Q.S.C.&V. And today
the concept is as alive and well as ever—with 7,000 restaurants,
and 40 billion hamburgers sold to date in a $2.5 billion enterprise.
On page 4 of the 1980 McDonald's annual report, the page after
the obligatory letter to shareholders, the first sentence starts:
"Quality is the first word in the McDonald's motto of Q.S.C.&V. . . .
That's because quality is what consumers enjoy each time they visit
a McDonald's restaurant."

"Sure," says the cynic, "don't all companies talk like that?" In
one of many checks on the McDonald's story, we interviewed a
friend, now a young business executive, who worked at McDonald's
as a seventeen-year-old high-school student. The interview was purposely unstructured, so that he could say whatever he wanted. He
quickly got around to quality, service, and cleanliness. "What impressed me, looking back on it," he said, "was the quality of ingredients. McDonald's always uses prime beef—the best of everything." He continued, "If french fries were overdone, we threw
them out . . . if we punched holes in the buns with our thumbs [a
frequent occurrence, especially for those new at the tough job of
handling thousands of buns], we threw them out. The incredible
thing to me is that here I am, thirteen years later, and if I want fast
food, I *still* go to McDonald's. French fries were their best product,
I always thought." (He's in good company. Julia Child loves Mc-
Donald's french fries, too.)

McDonald's is equally fanatical about cleanliness. Talk to an ex-
employee about what he or she remembers best, and almost invariably they'll tell you about constantly cleaning. "There was never an
idle moment," recalls one former griddle-tender. "Whenever there
was slack time in the store, we were cleaning something."

The mundane stories of consistent product and service from ex-griddle tenders are reinforced by brilliant strategic thinkers. Donald Smith, now a senior PepsiCo executive, left McDonald's to head up arch-competitor Burger King a few years ago. It's interesting to note that Smith set as his number one strategic priority making Burger King "more consistent [in appearance and service] across the country." He made a big dent in the problem in five years at the helm. But tackling McDonald's is tough. Smith's successor at Burger King, Jerome Ruenheck, is still hammering on the same theme. "The problem is consistency. They're more consistent than we are across the country."

In the original excellent companies research and in our continuing pursuit of the matter, we repeatedly found examples of those who pursue quality with quixotic zeal. *Digital* clearly falls into that category. The corporate philosophy states that "growth is not our principal goal. Our goal is to be a quality organization and do a quality job, which means that we will be proud of our work and our products for years to come. As we achieve quality, growth comes as a result." The paramount objective at *Maytag,* to repeat another example, is "Ten years' trouble-free operation" for any machine. At this late point in the product life cycle, washing machines should be almost commodities, like wheat or potato chips. Yet Maytag's devotion to dependability earns a full 15 percent price premium, while top share is maintained against determined competitors such as GE. Quality and reliability are, in fact, a life raft for all points in the economic cycle. While GE suffered severe recessionary blues at Louisville in its home laundry business, and while all appliance makers strained to survive, Maytag profits again grew, although not with boom-time vigor. Maytag's form of quality does not come from exotic technology; it comes from products that work. An analyst observes: "Maytag built its reputation on solid dependability, not jazziness. . . . It makes things good and simple."

The examples continue. At *Holiday Inns,* reliability is a paramount institutional goal and the fundamental theme of "no surprises" pervades the institution—including its advertising. *Procter & Gamble* believes deeply in the quality of the products it produces, so deeply that one analyst calls it P&G's "sometimes Achilles' heel." For example, P&G usually won't match competitors on trendy features. "P&G is weakest when it tries to respond to competitors that offer superficial, cosmetic advantages, such as flavor, rather than superior performance such as cavity-prevention," says an observer. "Cosmetics don't sit well with the Calvinists at Sixth and Sycamore," the address of P&G corporate headquarters in Cincinnati.

A vignette from a young ex-brand manager who had the Charmin brand of toilet paper illustrates the overwhelmingly positive side of P&G's reverence for quality. He was describing how customer complaints get sent back directly to the brand manager for action, and he recalled one intriguing incident. There are, it seems, three kinds of toilet paper dispensers: the kind you find in public washrooms, the kind that is typically mounted on the wall at home, and an old-fashioned type that is half built into the wall and fits into a semicylindrical wall cavity. It turns out that a roll of Charmin is about an eighth of an inch too thick to fit into the old-fashioned type. P&G's solution was most emphatically not to cut back on the number of sheets of paper, thereby compromising quality. Instead, the engineering department, R&D, and the brand manager got together and came up with an idea for tooling a machine so that it would wind the toilet paper faster, thereby reducing the diameter of the roll enough to fit into the dispenser.

Hewlett-Packard's Computer Systems Division makes the HP 3000. The system, first sold in 1968, was installed in 5,000 locations throughout the world by 1980, and is installed in over 8,000 sites today. The system typically ranks right at the top in terms of quality, as measured by a variety of independent outside surveys. It seems strange, then, that flush with sales and quality success, the Computer Systems Division would have undertaken a major new quality program last year for the HP 3000. But that is exactly what they did. Their all-too-rare attitude is: "If we don't keep up our quality momentum, the Japanese will leapfrog our position."

What immediately strikes one about the division's current quality program is the zealotry and the extent to which it pervades the whole operation. Although we hardly need to say it by now, the attitude starts at the top. Richard Anderson, the division manager, spends every fourth week in the field, visiting installations, talking with customers, and attending sales meetings. In the process, he is inundated with first-hand data on customer needs and competitive moves. As part of this effort he specifically solicits feedback on quality.

Anderson introduced the latest quality campaign a year ago. He made the announcement, as is typical with major new HP programs, at "morning coffee break" in the cafeteria, where most of the division's 1,400 employees meet every week to talk about business. He asked his staff to start defining and measuring quality. He used the Japanese encroachment in the industry as an example and a reason for urgency. And as the year progressed, a variety of quality programs permeated the division.

By the end of the first year, as measured by such vital standards as mean-time-between-failures, the already superb quality had improved fully 100 percent. Anderson is shooting for another 100 percent improvement this year from a basis that already beats the industry by a wide margin.

Division management signaled early and dramatically that the quality drive was real. During one notable "morning coffee break," five pallets of defective circuit boards were hauled in and dumped on the floor. Management explained to the astonished onlookers that those boards, and some less visible software bugs, were the equivalent of $250,000 in lost profit sharing (most employees at HP are stock owners and part of the profit-sharing program). This act was to characterize the way the division punishes and rewards performance. For quality failures, everyone shares the blame. For achievements, individuals are singled out.

The quality program is rich in formal and informal rewards, starting with the simplest of all: management wanders about, giving individuals compliments. Ceremony marks quality recognition at the coffee meetings, team dinners, and division beer busts. At its most formal level, the group executive vice president held a 1981 award ceremony, again at a coffee meeting. The winners were the people who best met the quality objectives in their work areas. They received special plaques, pen sets, and free dinners. Their names were posted in the division lobby, and they won free trips to any other HP division seminar or sales office located in the United States. "Yes, that includes Hawaii," noted an HP manager.

The routine HP systems are made to reinforce the quality objective. Quality objectives are built directly into the MBO program—a program that everyone takes seriously at HP. Feedback is frequent. For example, every week the division manager tells everyone the latest quality figures along with the latest figures on shipments, sales, and profits.

Each department within the division is a part of a quality web. LACE at HP stands for Lab Awareness of Customer Environment. In that program, HP customers make presentations to the engineers on their own needs and reactions to HP products and services. According to one observer, "The meetings are always standing room only." In another program, software engineers take a stint at manning the telephones used by the sales representatives and also visit user sites to get direct customer advice. Most importantly, the quality assurance department is part of the development team. This is very different from most companies where the quality controls people are the bad guys—the cops—and are typically in conflict with the rest of the division.

Glooper Troopers, Quality Enforcers, Vintage Quality, Solution Squad—those are some of the names of the quality teams, HP's version of quality circles, now at work on HP's quality program. Today, HP management systems are full of quality objectives and measures and no department is exempt from the program. One observer said it perfectly: "A quality focus is ubiquitous at HP because the employees don't seem to be able to separate it from anything else they are doing. If you ask them about personnel, they talk quality. If you ask them about field sales, they talk quality. If you ask them about management-by-objectives, they talk about quality-by-objectives."

Quality and reliability are *not* synonymous with exotic technology. It was especially interesting, and surprising, to us to find that even in higher technology businesses, reliability was always preferred over sheer technical wizardry. The star performers consciously sacrifice an unproven technology for something that works. We call the phenomenon "second to the marketplace and proud of it." Here are a few characteristic examples:

Hewlett-Packard (again): "The company is seldom first into the market with its new products—Xerox and IBM, for example, were first with high-priced laser printers. The company's marketing strategy is normally that of a counterpuncher. A competitor's new product comes on the market and HP engineers, when making service calls on HP equipment, ask their customers what they like or dislike about the new product, what features the customer would like to have. . . . And pretty soon HP salesmen are calling on customers again with a new product that answers their needs and wants. The result: happy and loyal customers." (*Forbes*)

Digital: "We must provide reliability. We purposefully lag the state of the art by two or three years. We let our lead users—for example, government research labs—push us. Then we develop a reliable product for our OEM [original equipment manufacturer] customers and other end users." (Interview)

Schlumberger: "While sometimes a competitor will be first with a given item, when Schlumberger introduces the product it will be more complete and of better quality." (*Dun's Review*)

IBM: Going back to its early days, IBM has seldom put products on the market that are right in the forefront of new technology. UNIVAC and others have all showed the way; IBM has learned from others' mistakes. "It was rarely the first to take a new technical step, but it wasn't far behind. And time after time, its new lines were better designed and more effectively sold and serviced than those of competitors." (*Financial World*)

Caterpillar: Even in the world of less arcane technology, we find the same phenomenon. "Caterpillar is rarely the first to come up with a new offering in its markets. But being on the leading edge has never been one of the company's goals. It has built its reputation by letting other companies go through the trial and error process of introducing new products. Caterpillar later jumps in with the most trouble-free product on the market. Indeed, Caterpillar products do not usually sport the lowest price tag. The company relies, instead, on quality and reliable service to woo customers." (*Business Week*)

Deere: Deere is the clear top performer in farm equipment. It is to farm equipment what Caterpillar is to construction equipment. "Deere isn't saying whether it will market a rotary combine. 'My guess,' one security analyst says, 'is that Deere will come out with its rotary combine within two years . . . and try to benefit from competitors' early mistakes.'" (*The Wall Street Journal*)

Now, the excellent companies' apparent satisfaction with being second shouldn't fool anyone about their technical ability. Many of the excellent companies, such as HP, IBM, and P&G, are among industry leaders in basic R&D expenditures. What distinguishes them is that their bias is toward making technology work for the consuming public. New products that make it through *their* screens are, above all, aimed at consumer needs.

A counter, and all too common, strategy was described to us by a computer peripherals executive: "We rushed to the market with a new product, because it was a clearly superior technical device. We wanted to grab market share quickly. But reliability was awful. Our share peaked at fourteen percent and is now down below eight percent, while we should have had thirty or thirty-five percent of the market. A six-month delay in introduction to iron out the bugs would have done it. Damn it."

Some people who have listened to us harp on service, quality, and reliability wonder whether it is not possible to overdo the overkill. The answer is quite obviously yes. As Freddy Heineken puts it, "I must keep telling my marketing people not to make the [beer] bottle too elaborate with gold foil or fancy labels. Otherwise, the housewife will be too intimidated to take it off the supermarket shelf." One long-time student of the airline industry puts the same point this way: "Braniff thought quality meant Alexander Calder paint jobs and comely stewardesses. Delta knows it means planes that arrive on time." The answer to how much service is enough or what kind of quality is right lies in the marketplace. A friend states the case nicely in terms everyone can understand. "The customer who is looking for a seventy-five-cent salad doesn't expect avocados, but she does expect the lettuce to be crisp. The seventy-five-cent salad producer should concentrate on crisp lettuce and forget about finding cheap avocados."

Owing to good luck, or maybe even good sense, those companies that emphasize quality, reliability, and service have chosen the *only* area where it is readily possible to generate excitement in the average down-the-line employee. They give people pride in what they do. They make it possible to love the product. In *The Decline and Fall of the British Manager*, Alistair Mant (one more ex-IBMer) provides a nice, concrete illustration of the mechanics of instilling love of and care for products:

On the surface there is nothing particularly interesting about Platt Clothiers Ltd except its success. Beneath the surface it is an efficient and tightly controlled beehive of activity where everyone lives and thinks *overcoats*. If you ask Monty Platt about his sales and marketing organization, he will reply, "My overcoats sell my overcoats." At 11:00 A.M. each morning a bell sounds, and anyone who wants to moves into the design office to have a look at yesterday's production. A random sample of overcoats is there to touch, try on, pull apart, mull over; and the boss is there talking overcoats to his dispatch manager, junior production staff and his designers. *Monty Platt has managed to instill his enthusiasm for overcoats into everyone who works for him* [italics ours]. Of course, he has to talk about "marketing," "personnel," "production," and other such rarefied ideas, but no one there can be in any doubt about their essential context—overcoats. His relationship with employees is about *work* and for them it is about working for an outfit that knows what it is doing, cares about *that* and does it well. What is the moral? Not all firms have the luxury of a single-product orientation nor the coziness of an integrated, compact organization. But all firms make *something* and they differ wildly in how much care goes into *that*. If only they so organized themselves that the kinds of people with a feel for manufacturing, for making things and making them well, ended up in key positions of authority, the chances are that the whole situation would *feel* different. Such people have *integrity* in the precise sense of that word in a manufacturing system, and they generate a feeling of integrity, in the broader sense, all around them.

The impossible becomes almost possible in the excellent companies. Is a 100 percent quality or service program plausible? Most would guffaw at the thought. But the answer is yes and no. Statistically, it's no. In a big company, the law of large numbers ensures that there will be defects and breached service standards now and again. On the other hand, a friend at American Express reminds us, "If you don't shoot for one hundred percent, you are tolerating mistakes. You'll get what you ask for." Thus it is possible to be genuinely aggrieved at failure, *any* failure, despite the volume. Freddy Heineken says bluntly, "I consider a bad bottle of Heineken to be a personal insult to me." Mars, Inc. (the giant candy company), a very successful company in a highly competitive market, thrives on quality. A Mars executive offers this glimpse of Forrest Mars: "He is given to fits of unbridled rage, such as a time he discovered an improperly wrapped batch of candy bars and hurled the entire inventory, one by one, at a glass panel in a boardroom while frightened aides looked on." J. Willard Marriott, Sr., at eighty-two is still incensed at any sign of carelessness in a Marriott facility; until recently he read every customer complaint card.

True service- and quality-oriented companies can and do expect to get things right. There is a lot to be said for blind faith (coupled with elbow grease), for only with such a vigorous belief is the organization likely to pull together. When an IBM computer crashes, a Cat customer needs a part, a Frito sales manager needs more stock, or HP feels threatened by the Japanese, there is no issue. The organization brings all the resources it can muster to bear on the problem. But even with high standards, companies can get lax if just an occasional failure in quality and service is considered tolerable. A Digital executive summarizes: "It's the difference between day and night. One is the mind set that says, 'Doing it right is the only way.' The other treats the customer as a statistic. Do *you* want to be part of the population struck by 'failure within tolerance'?"

Economists talk about "barriers to entry," what it takes to compete in an industry. As is so often the case, the rational model leads us to get "hard" and "soft" mixed up on this one, too. We usually think of principal barriers to entry as concrete and metal—the investment cost of building the bellwether plant capacity addition. We have come to think, on the basis of the excellent companies data, however, that that's usually dead wrong. *The real barriers to entry are the 75-year investment in getting hundreds of thousands to live service, quality, and customer problem solving at IBM, or the 150-year investment in quality at P&G.* These are the truly insuperable "barriers to entry," based on people capital tied up in ironclad traditions of service, reliability, and quality.

NICHEMANSHIP

The customer orientation is by definition a way of "tailoring"—a way of finding a particular niche where you are better at something than anybody else. A very large share of the companies we looked at are superb at dividing their customer base into numerous segments so they can provide tailored products and service. In doing so, of course, they take their products out of the commodity category, and then they charge more for them. Take Bloomingdale's. The heart of its success is the boutique, and each boutique is tailored to a unique service or a modest-sized set of customers. Bloomingdale's parent, Federated Stores, follows the same strategy with Bullock's, I. Magnin, Rich's, and Filene's. "Each department is a separate showplace," observes one executive. Chesebrough-Pond's offers a nice example of reaching the top through tailoring. *Forbes* recently described chairman Ralph Ward's strategy this way: "Though he can play the mega-buck promotion game, he would just as soon catch a competitor napping [in a small market]." In 1978, for example, he launched Rave, aimed at the then $40 million-a-year home permanent market, which was dominated by Gillette's Toni. Ward says, "The category had been asleep for years. We introduced a product with no ammonia—and no smell—and now it is a $100 million-a-year market." Moreover, in a rare strategy for a consumer goods company, he makes product divisions independent in order to speed their further search for additional niches.

3M is the classic player at this game. Its chairman Lew Lehr says, "Our organization does not believe in making just a few bets. Our people make hundreds of little bets in the form of new products for specialized markets." Here is but one example. We recently talked to the chief executive of a $50 million Richmond, Virginia, printing company. It is a leader in high-volume offset printing, a modest-sized niche served by a variety of 3M products. 3M decided it wanted to really learn to do business with the segment the company represented. It attacked in full strength. Sales teams descended from St. Paul, chockablock with engineers and technical people, to try to figure out their problems. Then they invited the chief executive and some of his principals to St. Paul to lecture in several divisions on how 3M could best serve them. What we found especially attractive about the incident was not only the intensity of the 3M approach but also the flexibility. 3M teams from various product areas all responded to the opportunity. There were no turf battles and no bureaucratic delays. While the 3M magic goes much deeper, as we will see, the attitude that it will attack any niche, no matter what size, is astonishing.

Such examples lead one to wonder whether it is possible to oversegment. Theoretically—as is the case with service and quality—the answer is yes. In practical terms, though, maybe not. It seems to us that 3M, Digital, HP, and many other excellent companies have purposely allowed much more than normal proliferation. They over-slice the pie, according to normal marketing conventions; yet among the corporate giants their performance stands out. Nichemanship is not always tidy. But it works.

We find five fundamental attributes of those companies that are close to the customer through niche strategies: (1) astute technology manipulation, (2) pricing skill, (3) better segmenting, (4) a problem-solving orientation, and (5) a willingness to spend in order to discriminate.

MIT's James Utterback, a long-time student of the technology diffusion process, argues convincingly that "the new technology enters through a specialized market niche, a high performance use where you can bear those high costs." That's the way companies like Digital and even IBM seem to look at life. Remember the example of lead users pushing Digital into the new stages of technology? Where does Digital put its best sales engineers? On academic and big government lab accounts. In developing solutions for those customers, Digital often evolves the next generation for its more average user. The niche people are masters at learning about sophisticated technology in one niche, testing it with later users, ironing the bugs out, and passing that technology along to still others.

Niche people are also superb at pricing mainly on a value basis. They get in early, charge a lot for providing a tailored product to a discrete population, and, when and if other competitors come in, they get out. One 3M executive put it this way:

Our objective is first and foremost to have a steady flow of new products. Then, once we hit, we expect to dominate the niche, sometimes for only three or four years. During that stretch we price according to full value to the customer. We're providing a new tool that is some kind of a labor saver and we expect the market to pay what it's worth. Sure, we create an umbrella. But when others come in with approximations, perhaps at lower cost, rather than fight them for share, we usually give in—that is, get out. Because by then we're developing the next several generations of products for that market and others.

David Packard once reminded his managers of the source of a rare HP failure, in early hand-held calculator marketing: "Somewhere, we got the idea that market share was an objective," he said. "I hope that is straightened out. Anyone can build market share; if you set your price low enough you can have the whole damn market. But I'll tell you it won't get you anywhere around here."

Most banks have discovered that high-net-worth (i.e., wealthy) individuals are a very desirable segment. But most banks are still wondering how to get their high-net-worth programs under way, because in general they are held back by the major disadvantage of not understanding segment appeals. A nice exception is the following report by one bank executive:

We decided to go hard after the high-net-worth individuals. One good contact point seemed to be their accountants. So we went out and made a presentation to the partners in each of the Big Eight firms in [a major metropolitan area]. In seven of the eight firms, we were the first bank ever to make a presentation at their offices! In *every* case, we were the first bank ever to bring along senior officers to the session. It started to pay off immediately. In all eight cases, we got our first new business within one day of the presentation. In several instances, it started on the spot.

Nichemanship is frequently accompanied by a problem-solving mentality. IBM trains its salesmen not to be salesmen but *customer problem solvers*. 3M has always done the same. A General Instruments sales executive captures the spirit of getting to know the customer well enough to really solve his problems:

I remember my first job. I spent forever getting to know a small handful of customers really well. It paid off handsomely. I came in at 195% of quota, tops in my division. A fellow at corporate called me and said, "Good job, to be sure, but you average 1.2 sales calls a day and the company averages 4.6. Just think of what you could sell if you could get your average up to par." You can guess my response, after I came down off the ceiling; I said, "Just think what the rest could sell if they could get their calls *down* to 1.2"

The niche people are willing to spend to discriminate. Says Edward Finkelstein of Macy's, "So long as you spend what's necessary to make a store attractive, you'll prosper." For Finkelstein, that has meant spending extravagantly on boutiques to match the Bloomingdale's effort in New York. He's been successful as a result. A successful catalogue house like Fingerhut overspends on data accumulation. "Using our data better, we can open a personal store for each customer," an executive says. At Ore-Ida, the story is the same. They are misers when it comes to overhead spending. But when it comes to market testing, the budget is gold-plated. For years Ore-Ida has been invincible in frozen consumer potato products.

HOW COST-ORIENTED ARE THEY?

When we started our survey, we expected to find the excellent companies putting stress on cost or technology or markets or niches. In other words, we felt some would have strategies oriented to one thing, and some to another, but we weren't expecting any particular bias. But that is not what we found. While there are differences among industries, we did find a striking commonality: the excellent companies tend to be more driven by close-to-the-customer attributes than by either technology or cost.

As a way of illustrating this, we have taken fifty top-performing companies and arrayed them by industry and by what seems to be their dominant bias. Some observers will argue with one assessment or another. Moreover, no company ignores cost or technology altogether. But a strong emphasis on one variable does seem to come at the expense of attention to others. As the figure on the opposite page illustrates, we find high-performing companies in different industries to be mainly oriented to the value, rather than the cost, side of the profitability equation. We divided the companies up into separate industry segments—high technology, consumer goods, service, miscellaneous manufacturers, project management, and commodity. A brief discussion of each category seems useful.

In the high-tech category, only four of fourteen companies seemed to us to be largely or primarily cost-driven. These are TI, Data General, National Semiconductor, and Emerson. Of these, all but Emerson seem to have had trouble in the last several years and are reassessing their strategies. Data General and National Semiconductor have both agreed that the strategy for the future is likely to be in finding niches. Data General as a case is especially instructive. The company attempted to beat pioneer Digital at its own game. Data General focused on the OEM market and developed a strategy based on a small number of low-cost products. Along the

Cost		Service/quality/reliability		High-value-added nichemanship
HIGH TECHNOLOGY (14)				
Data General Emerson National Semiconductor Texas Instruments		Allen-Bradley International Business Machines Lanier		Digital Equipment Corporation Hewlett-Packard Raychem ROLM Schlumberger Tandem Wang
CONSUMER GOODS (11)				
	Blue Bell	Frito-Lay Mars Maytag Procter & Gamble	Avon Chesebrough-Pond's Fingerhut Johnson & Johnson Levi Strauss Tupperware	
SERVICE (12)				
	K mart	American Airlines Disney Productions Marriott McDonald's	Delta Ogilvy & Mather Wal-Mart	Bloomingdale's Citibank Morgan Bank Nieman-Marcus
MISCELLANEOUS MANUFACTURERS (4)				
Dana		Caterpillar Deere		Minnesota Mining and Manufacturing (3M)
PROJECT MANAGEMENT (3)				
		Bechtel Boeing Fluor		
COMMODITY (6)				
Amoco Arco Exxon	Dow			Du Pont Nucor Steel

way it developed, and even encouraged, an image of being the "tough bastards." A lead *Fortune* article in 1979 questioned Digital's product proliferation (which was bound to cause high cost) and its noncommission sales force, contrasting it with Data General's aggressive, high-commission sales force. But the worm turned. Digital moved away from OEM dependence, and along with Wang, HP, and Prime, led the way in providing user-friendly, flexible products. The policies that produced product overlap and a problem-solving sales force at Digital paid off. On the other hand, Data General's "bastard" image has hurt, delaying for a while at least its remarkable performance march.

Compared to its truly superb performance of the last couple of decades, TI has been struggling a little in the last few years and has turned its sights outward again, toward marketing. Its previous out-and-out obsession with cost and market share seems to us a major reason why that company missed out on recent generations of semiconductor leadership, has had trouble with the home computer, and has never gotten fully on track in consumer electronics. In chips, for example, while much of the organization's brain power turned to the problem of racing down the cost curve to set the industry standard in the 8K RAM, attention almost inadvertently turned away from future, larger RAM chips. That is the crux of the matter. Too much attention to cost causes an internal shift in attention that seeps in slowly, almost unobserved. In consumer goods, such as watches and calculators, TI's approach was again low-cost: "Make commodities and make ours the cheapest," they seemed to think. TI's consumer project not only has come a cropper against the Japanese, but it also seems to have directly drained key resources from critical chip innovation.

As we've already mentioned, both Lanier and IBM typify the companies that go overboard on service among the high technology bunch. True, IBM's labs might work many generations ahead with, for example, the Josephson junction, but its day-to-day products typically lag what is technically possible. Allen-Bradley, the conservative, privately held billion-dollar Milwaukee manufacturing controls company, also comes into the service, quality, and reliability part of the spectrum. The whole company resonates to the tune of quality and reliability; that is what counts in control systems.

One could make a case for listing others, especially HP and Digital, as service and quality-biased, but they and the rest of our top-performing high tech companies seem to be niche people above all. All are a crucible of small, entrepreneurial activities aimed at firing new products into the marketplace. Wang, for example, put more than one new product per week into the market in 1980. Wang's R&D hit rate, largely owing to the intensity of user connections, is said to be over 75 percent—a truly remarkable performance.

ROLM is a close parallel. It is remarkably user-oriented and is not a technology leader. ROLM has badly thumped AT&T's Western Electric in the PBX business simply by being better at tailored problem solving. Tandem's "Non-Stop Computer" is classically niche-aimed. "Every customer is a separate segment" is the Tandem watchword. Raychem sells complicated "smart" electrical connectors. It has long overspent on the training and development of its salesmen, and the reason is simple. Raychem salesmen are practical, day-to-day applications engineers. They sell their connectors on the basis of the high economic value of the product to the customer. Installation of a connector is labor-intensive, and tailor-made devices can cut labor charges extensively. The connectors are a microscopic fraction of the value of the eventual product—for example, large aircraft; therefore, the customer can, in fact, afford to pay a bundle. The Schlumberger story is much the same. Its cadre of 2,000 field engineers provide oil well logging (measurement) and other services to drillers. They are just like Raychem. Their service product is a tiny fraction of the total cost of oil-field operations, but the value to the user of Schlumberger's doing it right is enormous.

To the degree that our tale of high technology leaders represents a pattern, it should surprise us all. The so-called high tech companies are not, first and foremost, the leaders in technology. They are in high tech businesses, but their main attribute is reliable, high-value added products and services for their customers.

In *consumer goods,* for the purposes of this analysis, we have arrayed eleven companies. None, in our opinion, is primarily a low-cost producer. What they offer, instead, are service, quality, and reliability. The casual observer of P&G would tell you that the company succeeds because of advertising and brand management. The insiders will tell you it's their quality and testing fetish; when they have a rare problem, as they did with Rely tampons and toxic shock, they will move fast, with lots of dollars, to do their best to reinstate their reputation for quality. Frito-Lay is a clear winner in service. Maytag does it with reliability. Its long-running ad featuring "Old Lonely," the sad-faced Maytag repairman with nothing to do, says it all. Mars also falls cleanly into this category.

Lots of companies sell door-to-door, but nobody does it with the same intensity as Avon or Tupperware. We put these top performers in the high value-added, niche category simply because they go out and create their own markets.

Two clear leaders in the apparel industry are Levi Strauss and Blue Bell, yet, interestingly, they take different approaches. Levi's was founded on the quality principle and sticks with it, while much of its remarkable recent growth has come from special marketing acumen, suggesting that it has moved toward the niche end of the spectrum. Blue Bell, number two in the industry, has done remarkably well with a very strong cost orientation that complements its quality obsession.

J&J, in our minds, is uniquely a niche company. The firm consists of about 150 nearly independent companies, each of which has a primary responsibility to get new products out the door. J&J *lives* by a credo in which customers come first, employees second, the community third, and shareholders fourth and last. Chesebrough-Pond's stars at the same game.

Fingerhut, a major retail cataloguer, is a seemingly strange member of the set, yet it may be the ultimate niche company. Because of a remarkable system for tracking customers and customer profitability, virtually every individual customer is a separate market segment. For example, as *Fortune* notes, "A month before your son turns eight, you'll receive a packet including a personalized letter promising that if you'll agree to try any of the products offered, Fingerhut will send along a free birthday gift suitable for an eight-year-old boy. The more orders you place the more packets you receive in the mail. . . . Fingerhut concentrates on its core customers, including such service as 'pre-approved' credit in the midst of a recession when J. C. Penney and Sears were cutting back." Looking beneath the surface, we don't find magic in Fingerhut. It's not very sophisticated; it's just that no other major catalogue merchandiser has bothered to do it.

Next come a dozen *service companies.* At Ogilvy & Mather, for example, David Ogilvy insists that the agency live up to his dictum that the number one objective be unparalleled client service, not profitability. At Marriott Hotels, J. Willard Marriott, Sr., at eighty-two, is the same demon about quality that he was forty years ago. His son, who now runs the company, has picked up the same theme, and even Marriott's advertising focuses on the personal visits Bill Marriott, Jr., makes to all locations. In the airline business, Delta and American are at the top of any list of big top performers. They also top the list on service. American unfailingly comes out first in the consumer service surveys. Delta would do the same if the analysis focused on the niche the company chooses to serve— principally the business customer.

In banking, we have listed two prime examples: Morgan and Citibank. The banking industry today talks incessantly of building management skills to serve the large corporate customer; Morgan wrote the book on it decades ago. Citibank was the first major bank to realign its entire organization structure around market segments. They did that in 1970, and other banks are just starting to get to it.

The stars of the mass people-handling business are McDonald's and Disney. We have already discussed both. We find it nearly impossible to fault either on its ability to serve customers with consistent distinction and quality.

Who are the stars in retailing? Certainly Neiman-Marcus and Bloomingdale's stand out among the rest. When Neiman-Marcus opened in 1907, its first ad pronounced: "The Store of Quality and Superior Values." Bloomingdale's, as previously mentioned, exemplifies the niche approach.

Wal-Mart is the mass retailing success story of the late 1970s and early 1980s. And we return again to nichemanship and service. Since 1972, it has grown from 18 to 330 stores and from $45 million to $1.6 billion in sales. Among mass retailers it is the classic niche chain. It has done to K mart what Lanier has done to bigger competitors in stand-alone word processors. Wal-Mart puts "too many" stores in any one portion of its midwestern and southwestern turf. The reason is simple. In doing so, it discourages K mart from entry.

K mart deserves recognition as a top performer; but, like Emerson, the company is something of an anomaly. It has scored primarily by focusing on low cost. In fact, it is the only one of the twelve service companies that has cost as its prime distinguishing feature. In the process, though, it has not ignored quality. One might even argue that it has begun to take Sears's traditional place. "Value at a decent price" has been Sears's long-time philosophy, and it is quickly becoming K mart's.

In the *miscellaneous manufacturers category,* 3M is the niche exemplar: find a little market; penetrate it; get all you can out of it; move on to the next one. Caterpillar and Deere, whom we have also put into this catch-all category, are quality and reliability fanatics. They also have extraordinary relationships with their dealers. Finally, in this catch-all category, Dana stands out. Like Emerson, it has been mainly successful through an ability to keep cost down by constant productivity gains.

The undoubted *project management* stars are Fluor, Bechtel, and Boeing. Fluor and Bechtel are leaders in the large-project construction business. They both take pride in the quality and reliability of their services, and then charge a lot for both. Boeing worries about cost but mostly talks about the importance of quality and reliability. As we did the research, we found over and over again that the primary way to understand a company's orientation is to listen carefully to the way its people talk about themselves.

Finally, for the sake of completeness, we looked at a few of the stars in *commodity businesses.* Here, above all, being in a low-cost position is important. By definition, in commodity businesses, especially where you are selling to other companies and not to an end user, your cost position is crucial. (For example, a company like GE's subsidiary Utah International Incorporated makes enormous profits selling metallurgical coal to the Japanese. They do not gain a huge advantage from their marketing ability. They are simply the lowest-total-cost producer supplying quality coking and coal to Japanese steel producers.) Amoco, Arco, and Exxon are simply excellent operators and explorers. They can get oil out of the ground cheaper than the rest.

But even in commodity businesses, there are some interesting differences. Dow and Du Pont are a classic split, although both are winners. Dow, the basic upstream commodity producer, has clearly been the leader in recent years, because it had the right resource strategy—the low-cost strategy—when OPEC put an armlock on us. But, at least until the very recent past, Du Pont had the more enviable new-product record. Du Pont thrives on downstream innovation in the market niches that resulted from these new products.

The commodity steel business in general is not very profitable, but it has downstream exceptions, too. Nucor, a highly profitable company, thrives on higher value-added specialty-steel niches.

The foregoing analysis is hardly statistically valid. Nor do we conclude from it that cost doesn't count, or that, say, 80 or 90 percent of the best-run companies are overwhelmingly quality-, service-, or niche-oriented. However, we do think that the overall sample is a sound one, and we do think the data are sufficient to establish that for most top-performing companies something besides cost usually comes first. And that something is a special way of being close to the customer.

LISTENING TO THE USERS

The excellent companies are better listeners. They get a benefit from market closeness that for us was truly unexpected—unexpected, that is, until you think about it. Most of their real innovation comes from the market.

Proctor & Gamble was the first consumer goods company to put the toll-free 800 phone number on all its packaging. In its 1979 annual report, P&G says it got 200,000 calls on that 800 number, calls with customer ideas or complaints. P&G responded to every one of those calls and the calls were summarized monthly for board meetings. Insiders report that the 800 number is a major source of product improvement ideas.

There is surprising and powerful theoretical support for what P&G and others are doing. Eric von Hippel and James Utterback of MIT are long-time students of the innovation process. Not long ago Von Hippel looked carefully at the source of innovation in the scientific instruments business. His conclusions: of eleven "first of type" major inventions he looked at, *all* came from users; of sixty-six "major improvements," 85 percent came from users; of eighty-three "minor improvements," about two thirds came from users.

Von Hippel reports that not only did the ideas come from users; in the great majority of inventions he studied—including all the first of types*—the idea was first tested, prototyped, proved, and used by users, not by instrument manufacturers. Moreover, additional extensive pre-commercial diffusion of ideas was done by other users. That is to say, the lead user invented an instrument, built a prototype, and put it into service. Other savvy users picked it up. Only then did a producer get into the act, "performing process engineering work and enhancing reliability, while leaving the basic design and operating principles intact."

A group of Boeing executives provides directional support. They observe that, judging from their own experience, the von Hippel findings are extreme; they can think of a number of examples in which major ideas and prototypes came from their own development efforts. But they hasten to add that if the product is not immediately matched with a customer need and developed in full partnership with a customer, they drop it. "If we can't find an interested customer to work with us early on," one notes, "the idea is sure to be a loser."

The best companies are pushed around by their customers, and they love it. Who in Levi Strauss invented the original Levi's jean?

* Sophisticated instruments, such as the gas chromatograph, the nuclear resonance spectrometer, and the transmission electron microscope.

Nobody. In 1873, for $68 (the price for filing the patent application), Levi's obtained the right to market steel-riveted jeans from one of its users, Jacob Youphes, a Nevada buyer of Levi's denim. And, as already noted, Bloomingdale's invented faded jeans for Levi's. Almost all early IBM innovations, including the company's first computer, were developed in collaboration with the lead customer—the Census Bureau. When did 3M's Scotch Tape business take off? When a salesman, not the technical people, invented a handy desk-top dispenser for what had previously been a narrow-use industrial product.

And so it goes. Digital's edge? "They rely on customers to find uses for minicomputers, rather than burdening the company with huge costs of developing and marketing applications on its own. Digital salesmen, engineers selling to other engineers, nurture strong and lasting relationships with customers." The analyst who wrote that notes, "It's surprising how little they've caused their own growth. For years, they've been dragged along by interesting applications their customers came up with." Wang Labs' story is the same: "They will be more influenced by what the customer wants. Among other things they are planning to establish a joint research and development program in which the company will work along with its customers to determine new ways to use integrated systems." Founder An Wang says, "Working with users will help us respond to their needs." A top executive at Allen-Bradley notes, "We won't try anything unless we find a user who will cooperate with us in an experiment." He adds that Allen-Bradley had slipped behind in numerical control and programmable control devices. The company was subsequently pushed to the forefront not by its own researchers or engineers, but by its sophisticated lead users. "Boeing, Caterpillar, and GM were building their own equipment," said one executive. "They said in effect, 'Get on with it or forget us.'"

In one successful high technology company we talked with, the head of R&D has taken a two-month "summer vacation," as he calls it, in each of the last twelve years. In July and August he travels exclusively to user locations, carefully surveying what customers do with his company's products and what their future needs might be. We recently overheard a conversation in a Palo Alto bar. An HP engineer from an integrated circuit division was talking to friends. One asked him where he was working. He mentioned an HP site in Palo Alto, but quickly added that he spent almost all of his time working on applications in another city, on user premises.

These stories would be of little interest if they didn't stand in such marked contrast to most management practice. All too often the product is designed in a vacuum, the pipedream of engineers who love the technology but may never have seen living, breathing customers use their companies' products.

So the excellent companies are not only better on service, quality, reliability, and finding a niche. They are also better listeners. That is the other half of the close to the customer equation. The fact that these companies are so strong on quality, service, and the rest comes in large measure from paying attention to what customers want. From listening. From inviting the customer into the company. The customer is truly in a partnership with the effective companies, and vice versa.

Among the most extensive innovation studies have been the SAP-PHO* analyses, led by the noted economist Christopher Freeman. He analyzed thirty-nine innovations in the chemical industry and thirty-three in scientific instruments. More than 200 measures of aspects of innovation were used; only 15 proved to be statistically significant. The number one factor was the same in both industries: "Successful firms understand user needs better" (likelihood of respondents' mention of the factor occurring randomly rather than systematically: chemicals—.000061 (about 6/1000ths of one percent); instruments—.00195; combined—.00000019; i.e., it seems clearly valid). The number two factor was also the same for both, reliability: "Successful innovations have fewer problems." His analysis of specific failure cases was also revealing. The principal reasons mentioned by respondents were:

* Scientific Activity Predictor from Patterns with Heuristic Origin.

	Seven Failed Chemical Innovations	Sixteen Failed Instrument Innovations
"No inquiries of users at all"	1	3
"Too few inquiries or atypical users"	2	4
Ignored the (users') answers or misinterpreted	0	4
"No on-the-spot investigations of user techniques"	0	3
"Committed to preconceived design"	4	2

In summary, Freeman and his colleagues noted: "Successful firms pay more attention to the market than do failures. Successful innovators innovate in response to market needs, involve potential users in the development of the innovation, and understand user needs better."

We should not close this chapter without mentioning briefly a major debate that has taken place within our own ranks. It is our belief, based on the excellent companies review, that the user is supreme as a generator and tester of ideas. Several of our colleagues, on the other hand, maintain that companies are better driven by paying attention to technology and competitors. Moreover, Robert Hayes and William Abernathy, in a widely cited article in *Harvard Business Review,* have attacked U.S. companies for being too "market oriented" as opposed to "technology oriented." They argued that our short-term focus has led us to be captive to the latest consumer preference polls.

We disagree. First, we distrust any simple answer and aren't trying to push one ourselves. All three factors—users, competitors, technology—are essential. However, the competitor issue is easily put to rest. The excellent companies clearly do more and better competitor analysis than the rest. It's just that the work is not done by ivory tower staffers, reading or producing abstract reports. The

HP service rep, the IBM salesman, the 3M salesman or venture team leader, McDonald's franchisee, and Bloomingdale's buyer— by the hundreds or thousands, they are superb, intense competitor watchers. They do virtually all of it on location. And their sensor density is nothing short of overpowering.

The most controversial point our critics raise has to do with the technology issue; for example, "Users typically tell you more of the same, rather than suggesting true innovation." That may be true in some places (bulk commodity chemicals, for example), but not in many. Leaders in the sophisticated control business, such as Allen-Bradley, were driven to test robotics not by their central labs, but by their giant customers. IBM was really driven to distributed processing by its lead users, notably Citibank. NCR missed the electronics market in the late sixties by ignoring its lead users—Sears, J. C. Penney, *et al.*—and only recovered after forswearing its obstinacy.

The top "better listeners," then, pay especially close attention to their *lead users*. This is really the cutting edge, which differs greatly from the Hayes and Abernathy construct. The front-edge user (that is, the inventor rather than the average consumer), even in most consumer goods areas, is years ahead of the modal consumer, perhaps more than a decade in higher technology areas. (GM was a classic "lead user," ten years ahead of the herd in testing computer-aided design capability, which assisted the corporation immensely in beating Ford and Chrysler in world car design.) Similarly, little inventors can be found who are far ahead of the giant corporations in applications of new technology. And they, in turn, are working with others. Not surprisingly, there are a lot of such combinations at work at any given time. And the big company winners, our evidence strongly suggests, are those whose sales, marketing, manufacturing, engineering, and product development forces are close enough to their lead customers, and regularly enough in touch, to observe and quickly follow these user-innovator combinations.

Listening or sleuthing of this class, at or near the edge of the state of the art, is a long way from commissioning polls or convening panels to discuss yesterday's tastes. It's also a long way from the Hayes/Abernathy lab-based, pure technology play. Of course, one ought to invest in basic R&D. But its major role is arguably to spin out ideas that pragmatic internal entrepreneurs, such as champions, problem-solving salesmen, lead customers, and customer-oriented marketers, can "steal," fiddle with, and apply—today.

7

Autonomy and Entrepreneurship

The new idea either finds a champion or dies. . . . No ordinary involvement with a new idea provides the energy required to cope with the indifference and resistance that major technological change provokes. . . . Champions of new inventions display persistence and courage of heroic quality.

—Edward Schon, MIT

The most discouraging fact of big corporate life is the loss of what got them big in the first place: innovation. If big companies don't stop innovating entirely, the rate almost certainly goes way down. According to *Inc.,* a National Science Foundation study finds that "small firms produced about four times as many innovations per research and development dollar as medium-sized firms and about twenty-four times as many as large firms." Studying the same subject, the economist Burton Klein found that major firms are seldom if ever responsible for the major advances in their industries. Veronica Stolte-Heiskanen recently concluded a major study of fifty private and public sector research labs; her finding was roughly the same: "The relationship of objective material resources [fiscal and personnel] to research effectiveness is . . . generally minimal and sometimes negative."

On the other hand, we have the excellent companies. They are big. They have enviable records of growth, innovation, and consequent wealth. Clearly, the odds are stacked against them, yet they do it just the same. Perhaps the most important element of their enviable track record is an ability to be big and yet to act small at the same time. A concomitant essential apparently is that they encourage the entrepreneurial spirit among their people, because they push autonomy remarkably far down the line: Dana with its "store managers," 3M with its venture teams, TI with over ninety Product Customer Centers. At Emerson Electric and J&J we found "too many" divisions, and consequently a typical division size that at first looked suboptimal. Many of these companies were proud of their "skunk works," bands of eight or ten zealots off in the corner, often outproducing product development groups that numbered in the hundreds.

It eventually became clear that all of these companies were making a purposeful trade-off. They were creating almost radical decentralization and autonomy, with its attendant overlap, messiness around the edges, lack of coordination, internal competition, and somewhat chaotic conditions, in order to breed the entrepreneurial spirit. They had forsworn a measure of tidiness in order to achieve regular innovation.

But the more we looked, the more bewildering, in a sense, it all became. People talked of performance shootouts (IBM), killing programs at least once (3M), Fellows' and Individual Contributors' programs (IBM and TI), station managers (United Airlines), supporting failures (3M, J&J, Emerson), volunteering for critical projects, hiving off new divisions, finding listeners, bootlegging (GE), drilling more wells (Amoco), attacking simultaneously on many fronts (Bristol-Myers), and encouraging gadflies and mavericks (IBM). If we were not already convinced that the military metaphor was woefully inadequate to describe managerial life in the excellent companies, we certainly knew it after analyzing successful innovation schemes.

But we felt that there must be something more to it than just radically decentralizing and then urging the troops to "be creative, damn it," as a colleague describes the typical approach to trying to achieve innovation. And it turned out that there was.

THE CHAMPION

All the activity and apparent confusion we were observing revolves around fired-up "champions" and around making sure that the potential innovator, or champion, comes forward, grows, and flourishes—even to the extent of indulging a little madness. For, as Tait Elder, then head of the New Business Ventures Division at 3M, told us flatly, "We expect our champions to be irrational."

Howard Head is a champion par excellence. James Brian Quinn says of him and his revolutionary ski: "He was possessed by his idea, a fanatic on the subject." To understand what champions are really like, read *Sports Illustrated*'s story of Head's invention of the metal ski:

In 1946 Head went off to Stowe, Vt., for his first attempt at skiing. "I was humiliated and disgusted by how badly I skied," he recalls, "and, characteristically, I was inclined to blame it on the equipment, those long, clumsy hickory skis. On my way home I heard myself boasting to an Army officer beside me that I could make a better ski out of aircraft materials than could be made from wood."

Back at Martin, the cryptic doodles that began appearing on Head's drawing board inspired him to scavenge some aluminum from the plant scrap pile. In his off-hours he set up shop on the second floor of a converted stable in an alley near his one-room basement flat. His idea was to make a "metal sandwich" ski consisting of two layers of aluminum with plywood sidewalls and a center filling of honeycombed plastic.

Needing pressure and heat to fuse the materials together, Head concocted a process that would have made Rube Goldberg proud. To achieve the necessary pressure of 15 pounds per square inch, he put the ski mold into a huge rubber bag and then pumped the air out through a tube attached to an old refrigerator compressor that was hooked up backward to produce suction. For heat, he welded together an iron, coffin-like tank, filled it with motor oil drained from automobile crankcases and, using two Sears, Roebuck camp burners, cooked up a smelly 350° brew. Then he dumped the rubber bag with the ski mold inside into the tank of boiling oil and sat back like Julia Child waiting for her potato puffs to brown.

Six weeks later, out of the stench and smoke, Head produced his first six pairs of skis and raced off to Stowe to have them tested by the pros. To gauge the ski's camber, an instructor stuck the end of one into the snow and flexed it. It broke. So, eventually, did all six pairs. "Each time one of them broke," says Head, "something inside me snapped with it."

Instead of hanging up his rubber bag, Head quit Martin the day after New Year's 1948, took $6,000 in poker winnings he had stashed under his bed, and went to work in earnest. Each week he would send a new and improved pair of skis to Neil Robinson, a ski instructor in Bromley, Vt., for testing, and each week Robinson would send them back broken. "If I had known then that it would take 40 versions before the ski was any good, I might have given it up," says Head. "But fortunately, you get trapped into thinking the next design will be it."

Head wrestled with his obsession through three agonizing winters. The refinements were several: steel edges for necessary bite, a plywood core for added strength, and a plastic running surface for smoother, ice-free runs. One crisp day in 1950, Head stood in the bowl of Tuckerman's Ravine in New Hampshire and watched ski instructor Clif Taylor come skimming over the lip of the headwall, do a fishtail on the fall line and sweep into a long, graceful curve, swooshing to a stop in front of the beaming inventor.

"They're great, Mr. Head, just great," Taylor exclaimed. At that moment, Head says, "I knew deep inside I had it."

Recently, TI conducted a fascinating survey, reviewing its last fifty or so successful and unsuccessful new-product introductions, and found that one factor marked *every* failure: "Without exception, we found we hadn't had a *volunteer* champion. There was someone we had cajoled into taking on the task." The executive who told us this added: "When we take a look at a product and decide whether to push it or not these days, we've got a new set of criteria. Number one is the presence of a zealous, volunteer champion. After that comes market potential and project economics in a distant second and third."

In a parallel effort we recently finished an analysis of the last twenty years' performance of a dozen or so major U.S. and Japanese companies. One part was an in-depth study of twenty-four major business initiatives, such as GE's unsuccessful foray into computers and its success in engineered plastics and aircraft engines. Here again, the role of the champion proved crucial. In fifteen of the twenty-four cases that were successful, fourteen involved a clear champion, while of the nine failures, just three were champion-led. (Six either had no champion, or the champion had left early and the project consequently had fallen apart.) To our surprise, moreover, the Japanese and American data matched. We had expected few champions in the purportedly more collectivist Japanese environment. Yet 100 percent (six out of six) Japanese successes had a champion, and three of the four Japanese failures had none.

We'll admit that Head is central casting's idea of the prototypical inventor, working in a musty, smelly garage. But company men at Hitachi and GE? Yes, and at IBM too. James Brian Quinn, reviewing a quarter century of IBM history, says: "Committed champions were encouraged to carry forward major developments. Chairman Vincent Learson created this style at IBM during the company's most innovative period. He encouraged different groups to bring forward proposed designs for 'performance shoot-outs' against competing proposals. It was, in fact, difficult to find any successful major IBM innovation that derived directly from formal product planning rather than this championship process."

An ex-IBMer who was on hand in the senior Watson's years likewise notes: "The 650 [an early, critical IBM computer] was typical. The Poughkeepsie [central labs] bunch were moving along slowly. A group in Endicott [the manufacturing and engineering headquarters] had a simple little bootlegged project going. Armonk [headquarters] got wind of it. It was a hell of a lot better—simpler, cheaper—than the lab's product. It became the 650." Discussion with an IBM manager in San Jose added further confirmation:

Parallel projects are crucial. No doubt of it. When I look back over the last dozen products we've introduced, I find in well over half the instances the big development project that we "bet on" via the system came a cropper somewhere along the way. In every instance—and we've gone back and taken a look and I do mean *every*—there were two or three (about five once) other small projects, you know, four- to six-person groups, two people in one instance, who had been working on parallel technology or parallel development efforts. It had been with scrounged time and bodies. But that's a time-honored thing. We wink at it. It pays off. Looking at the projects where the initial bet failed, the subsequently developed project came in ahead of the original schedule in three instances. It's just amazing what a handful of dedicated people can do when they are really turned on. Of course, they had an advantage. Since they were so resource-constrained, they had to design a simpler product in the first place.

The GE story is the same. A look beneath the surface unearths tale after tale. One of GE's biggest recent commercial successes, other than acquisitions, for example, has been engineered plastics (from nothing in 1970 to $1 billion in 1980). The idea for engineered plastics came from off-line activity, says a *Dun's Review* commentator:

Like most companies, GE finds that some of its researchers' ideas do not have enough apparent promise even for the Schenectady lab [central R&D lab] to finance. Thus the company leaves enough maneuvering room for an ambitious researcher to engage in clandestine work that is surreptitiously financed by funds allocated for another project. Commonly known at GE as "bootlegging," such unauthorized research can sometimes pay big dividends. In the 1950s, a researcher named Daniel W. Fox, who was working on a new insulating material for electrical wiring, walked into Beuche's [the head of technology] office with a big glob of brown plastic on the end of a glass rod. Fox laid it down, hit it with a hammer, and the hammer broke. He tried to cut it with a knife, but couldn't. The material was demonstrated to the new chemical development unit, which refined it into a substance called Lexan polycarbonated plastic, creating what is now GE's fastest growing business.

It wasn't quite that simple. Fox, the technological champion, was not enough. Several other major actors were required to get it through the bureaucracy and successfully into the marketplace. Young Jack Welch, (now chairman) was the classic champion. He bootlegged incessantly, found niches in which to experiment with customers, went outside the system to recruit young chemical engineers who could develop Lexan further. Moreover, Welch himself was protected by a handful of strong, iconoclast "executive champions."

One may ask, if so many voices agree that champions are pivotal to the innovating process, why don't companies simply go out and hire or develop more of them? Part of the answer seems to be that the champion's working style is at odds with the way most businesses manage. We quote James Brian Quinn again:

Most corporations fail to tolerate the creative fanatic who has been the driving force behind most major innovations. Innovations, being far removed from the mainstream of the business, show little promise in the early stages of development. Moreover, the champion is obnoxious, impatient, egotistic, and perhaps a bit irrational in organizational terms. As a consequence, he is not hired. If hired, he is not promoted or rewarded. He is regarded as "not a serious person," "embarrassing," or "disruptive."

Another factor seems to be a certain confusion between creativity and innovation. Harvard's Theodore Levitt states the case as well as anyone else:

The trouble with much of the advice business gets today about the need to be more vigorously creative is that its advocates often fail to distinguish between creativity and innovation. Creativity is thinking up new things. Innovation is doing new things. . . . A powerful new idea can kick around unused in a company for years, not because its merits are not recognized, but because nobody has assumed the responsibility for converting it from words into action. Ideas are useless unless used. The proof of their value is only in their implementation. Until then, they are in limbo.

If you talk to the people who work for you, you'll discover that there is no shortage of creativity or creative people in American business. The shortage is of innovators. All too often, people believe that creativity automatically leads to innovation. It doesn't. Creative people tend to pass the responsibility for getting down to brass tacks to others. They are the bottleneck. They make none of the right kind of effort to help their ideas get a hearing and a try. . . .

The fact that you can put a dozen inexperienced people in a room and conduct a brainstorming session that produces exciting new ideas shows how little relative importance ideas themselves have. . . . Idea men constantly pepper everybody with proposals and memorandums that are just brief enough to get attention, to intrigue and sustain interest—but too short to include any responsible suggestions for implementation. The scarce people are the ones who have the know-how, energy, daring, and staying power to implement ideas. . . . Since business is a "get-things-done" institution, creativity without action-oriented follow-through is a barren form of behavior. In a sense, it is irresponsible.

A senior officer in a successful consumer goods company underscores Levitt's point with a very practical example. "The product winners are always championed," he says,

by a brand manager who has ventured far beyond the rules. He has worked with R&D on an intense, personal basis (most of his less successful cohorts worked only formally with researchers); as a result he garners an "unfair" share of R&D time and attention. Similarly, straying far beyond his official charter, he gets involved in a hands-on way with pilot manufacturing. All in all, his intensity leads him to try more things, learn faster, get lots more time and attention from other functions—and eventually to succeed. There is no magic. I can get five guys in R&D together any afternoon and come up with seventy-five to one-hundred plausible new product ideas. The point is to get on with testing and moving ahead. There are no geniuses in this business. You just gotta keep at it.

The champion is not a blue-sky dreamer, or an intellectual giant. The champion might even be an idea thief. But, above all, he's the pragmatic one who grabs onto someone else's theoretical construct if necessary and bullheadedly pushes it to fruition.

Championing Systems

In Chapter 5 we recounted the story of Sam Neaman. He was a true champion at McCrory's but not the only one. How about the fellow in Indianapolis who did his first demonstration store for him? In the case of GE's entry into engineered plastics, we unearthed several heroes: the inventor, the entrepreneur inside the company, and the executive champions who protected the others from the bureaucracy.

An author in *Research Management* recently concluded: "One-man shows are seldom effective. . . . Entrepreneurs often need a sponsor." The numerous schemes describing systems of championing all come down to the same thing—some form of primary champion plus some form of protector. As we move from consideration of the individual to the organization, we find there is a need for a number of players pushing innovation forward.

Our observations have led us to identify three primary roles: the product champion, the executive champion, and the godfather.* (We've intentionally left out the technical innovator, or inventor, because we don't view the initial technical work, the idea work, as a principal variable in innovating. The constraint on innovation, we believe, is almost always the absence of a product champion, executive champion, or godfather. Mostly, we are convinced of the importance of the executive champion and the godfather.)

The *product champion* is the zealot or fanatic in the ranks whom we have described as being not a typical administrative type. On the contrary, he is apt to be a loner, egotistical and cranky. But he *believes* in the specific product he has in mind.

The successful *executive champion* is invariably an ex-product champion. He's been there—been through the lengthy process of husbanding, seen what it takes to shield a potential practical new idea from the organization's formal tendency toward negation.

The *godfather* is typically an aging leader who provides the role

---

*We are not the first to propose such a scheme. MIT's Edward Roberts, Dartmouth's James Brian Quinn, and Stanford's Modesto Maidique, among others, have all proposed some form of hierarchy of champions.

model for championing. The mythology at 3M, HP, IBM, Digital, TI, McDonald's, and GE is crucial to the practical, lengthy process of product innovation. The myths of Lewis Lehr and Raymond Herzog *et al.* (3M), Edison, Welch *et al.* (GE), Hewlett (HP), Olsen (Digital), Wang (Wang), and Learson (IBM) are essential to fostering the plausibility that animates the overall championing system. A young engineer or marketer simply does not step out and take risks because of some "good feeling" in the gut. He steps out and takes risks because the history of the institution supports doing so as a way of life that leads to success. And he does so despite the certainty of repeated failure.

Playing the Numbers. Not surprisingly, most champions fail most of the time. If we state, then, that champions and systems of champions are the single most important key to sustained innovative success in the excellent companies, how do we reconcile repeated failure and overall success? Only one way: innovation success is a numbers game.

Now suppose that a new initiative is launched and its odds of succeeding are only 10 percent. If ten such initiatives are launched, the laws of probability tell us that the odds that at least one thing will work go all the way up to 65 percent. If twenty-five such initiatives are launched, the odds of at least one thing succeeding go up to more than 90 percent (the odds of at least two successes are almost 75 percent). The crystal-clear message is that no matter how small the odds are of any one thing's working, the probability of something's succeeding is very high if you try lots of things. According to James Brian Quinn, "Management must allow a sufficient number of projects with a long enough lead time for the characteristic 1:20 success ratio to have effect. Initially, entrepreneurial managers may need to undertake projects in somewhat lower risk ratios in order to build management confidence."

The only way of assuring more "hits" is to increase the number of "at bats." Thus Digital, HP, 3M, TI, Bloomingdale's, IBM, McDonald's, 3M, GE, Wang, J&J, and others simply have more would-be champions out there than their competitors do. Indeed,

Digital treats every customer, for all practical purposes, as a new product test site.

A recent analysis of Bristol-Myers's success provides a crisp example of success by the numbers. Richard Gelb has put together a superb record as Bristol's chairman. *Forbes* declares that Gelb is willing regularly to come in second: "Dick Gelb says, 'As long as we can put two second bests together, somehow we'll come out better. Runners up make more money.'" And *Forbes* observes that "Gelb attacks simultaneously on enough fronts that if any single product doesn't work out over time, he can cut his losses quickly." Bristol's numbers support the validity of Gelb's strategy. During the last five years, thirty-three health and beauty care products, in the general market, have been judged as commercial successes (annual food store sales of $5 million or more). According to *Forbes*, "Bristol-Myers had eight. The next best showing: three." Gelb comments: "Blockbusters are great, but there are other ways of moving ahead in the ethical drug business. We're not putting all of our eggs into the basket of one miracle drug for the future. If I've got $1 billion of pharmaceutical sales, I'd be a lot happier with 10 businesses at $100 million than two at $500 million." *Forbes* summarizes, "So Bristol hits fast, gets a lot of products out there, makes money now. Bristol's major strength is precisely that it *doesn't* pour $250 million into the research pipeline, sit back and pray that one of these days somebody will come up with a cure for cancer."

The numbers game is most obvious in a business like oil. Under chairman John Swearingen, Amoco, for instance, has achieved a record of domestic oil exploration success that is at the top of the industry—better than even, say, Exxon, Arco, or Shell. The mainspring is simple numbers. "Standard likes to drill as many wells as possible," notes a *Fortune* commentator. "The passion for exploration by whatever means are available sets Amoco sharply apart from the other majors. Exxon, for instance, rarely drills a well unless it owns the whole deal. And Amoco's George Galloway [head of production] was startled to learn at a recent briefing in Houston that Mobil was operating with just 500,000 leased acres in a region where Amoco had 20 times more." (Galloway adds, "Mobil must be pretty confident to pinpoint its exploration so minutely. I don't think we're that smart.")

The numbers story would hardly be worth telling were it not for the "home-run only" mentality that marks most businesses, even oil. The home-run mentality proceeds from a misplaced belief in planning, a misunderstanding of the disorderly innovative process, a misguided trust in large scale, and an inability to comprehend the management of organized chaos and lots of base hits.

Support for Champions. Champions are pioneers, and pioneers get shot at. The companies that get the most from champions, therefore, are those that have rich support networks so their pioneers will flourish. This point is so important it's hard to overstress. No support systems, no champions. No champions, no innovations.

What strikes us most about the excellent companies is the completeness of their support systems for champions. In fact, the excellent companies are structured to create champions. In particular, their systems are designed to "leak" so that scrounging champions can get something done.

They often do this in "skunk works." At a $5 billion survey company, for example, three of the last five new-product introductions have come from a classic skunk works. It consists at any one time of eight to ten people, and is located in a dingy second-floor loft six miles from the corporate headquarters. The technical genius is a fellow whose highest degree is a high-school equivalency diploma earned in the Army in Korea (although the company has literally thousands of Ph.D. scientists and engineers on its payroll). One of the other members of the group was arrested for sneaking into a manufacturing facility to which he had no pass and swiping some material needed to get on with an experiment.

The group's first product, now a $300 million per year sales item, was fully developed (prototyped) in twenty-eight days. Last year a major corporate product bombed. A skunk works member asked for and got permission to take two samples home and set them up in his basement. He used one as a benchmark. He tinkered with the other for about three weeks and corrected virtually all of the flaws (with nickel and dime items), actually improving performance over original design specs by a factor of three. The president visited his basement and approved design changes on the spot. The latest of the group's successes was designed in (covert) competition with a corporate engineering "team" of almost 700 people.

Skunk works are notoriously pragmatic, as is shown by yet another vignette about this group. A part of a major new machine was overheating. Big teams of engineers wrestled with the problem for months. Finally it was decided to mount a 1-ton air conditioner on the machine. One of the fellows from the skunk works happened by. He looked at the problem, then proceeded to the corner pharmacy to pick up an $8.95 household fan. It filled the bill, lowering the temperature sufficiently to fix the problem.

The places where we heard about skunk works tended to be those where more elaborate structures for supporting or encouraging champions did not exist. In the very top performing companies we heard more about something for which our colleague David Anderson devised the term "limited autonomy position"—meaning a position that has substantial entrepreneurial, champion-like qualities, but is actually quite constrained and exists in a much broader setting than one might expect.

We first ran across the concept in an analysis of United Airlines, when it was prospering under Ed Carlson's direction. Carlson talked of "simulated entrepreneurship." He gave some 1,900 "station managers" at United some control over their destinies. They were marked or graded for the first time not on total performance but on those variables over which they had some control. Says Carlson, "We were trying to present a realistic challenge to each station manager so that at the end of six months he could say to his boss or to his wife, 'I made a profit.'"

We next encountered the phenomenon at Dana, where chairman Rene McPherson invented the "store manager" concept, as mentioned before. Practically speaking, it meant giving his approximately ninety factory managers lots of authority. They had unusual control over hiring and firing; they had their own financial control systems; they did their own purchasing—all tasks that are normally centralized. McPherson's view is that these are the men on the front line; they're likely, over the long haul, to make better decisions than any central staff.

The same concept goes by the name "brand manager" at Procter & Gamble and Frito-Lay. The brand manager is, in reality, anything but a swashbuckling entrepreneur. On the other hand, the entire socialization process of, say, the P&G system is aimed at making him believe that that's exactly what he is, a hero. Time and again the system of myths and tales lauds the valiant brand manager who has challenged those many years his senior and repositioned his brand against all odds (and in competition with all the other brand managers).

At Schlumberger, in the oil equipment business, the simulated entrepreneurs are the 2,000 young oil-field engineers sent out to isolated locations, the ones of whom D. Euan Baird, head of logging (drilling log) operations, has said: "To me, Schlumberger is the fellow who goes out to the well a bit anxious, provides good answers for the customer, and drives away believing he's King Kong." The attrition rate is high. But they *are* Schlumberger, out where it counts—in the boondocks. The responsibility, by some measures, is quite limited. Yet, indeed, each is trained to believe he is truly empowered.

At IBM, Digital, and Raychem the limited autonomy position is salesman-as-problem-solver. Tom Watson launched the concept at IBM around 1920. Digital follows it today and calls the process of getting close to the customer "warm armpit marketing." 3M is known to outsiders as "the salesman's company." It got its start when its salesmen avoided the purchasing agents and went directly to the operators on the shop floor. The method is still practiced today by 3M's sales force. Raychem hires virtually all of its salesmen out of the Harvard Business School. They start out as salesmen and act as sophisticated problem solvers.

In our view, there is only one trick to making positions like this work. But it's a tough one. It involves *socializing* the managers to believe they are would-be champions, yet at the same time maintaining very substantial control where it counts. Most companies that can't think beyond such platitudes as "Authority must match responsibility" can't come to grips with this tough dual task. Many companies introduce brand- or product-management schemes; heaven knows how many have tried to copy P&G. But what they seldom take the time to do is create the mythology, role models, and structure of heroes that transfer the burden (commitment, fire) onto the brand managers. Or if they do transfer the burden onto their brand managers, as some do, then they don't play the other half of the game well—providing the incredibly tight and regular support systems that silently prop up the P&G brand manager and aid him in getting his job done. The P&G case is *the* classic. The brand manager is, on the one hand, taught that if he acts like King Kong in the marketplace, he may end up as chairman some day. Yet, with the discipline abetted by the vertical brand-management structure and the small number of "deeply grooved" systems, his autonomy is in reality extraordinarily limited. It's some juggling act.

"Suboptimal" Divisions. A $6 billion organization we encountered some years ago had organized technical groups into "competency centers"—physics, chemistry, et cetera. These centers had become the primary organizational elements. Projects and products ranked a distant second. The practical outcome of the imbalance was that an individual's time was hopelessly fragmented. Any person might work on as many as a half-dozen projects associated with his narrow specialty. The projects might, in turn, span three or four divisions, two or three groups. The organization was a disaster. Very little was delivered on time—principally, in our view, because of a lack of commitment and a focus on the wrong things, technical disciplines rather than products, projects, and customers. When the organization returned, after a five-year hiatus, to a project mode (with technical competency relegated to a distant second), development activities picked up noticeably—and almost overnight.

Contrast that with HP. The $3.5 billion company has fifty small divisions (average $70 million). Each division is limited to about 1,200 employees. One of us recently visited a division that had edged up to 2,000 people. Its natural solution: reorganize into three units, each, as usual, with full product development capability. Like 3M, the name of the game is not to grow bigger divisions, but to spawn, or "hive off," new ones. One commentator adds, "In carrying out its basic mission, an HP division conducts itself much like an independent business. As such, it is responsible for its own accounting, personnel activities, quality assurance, and support of its product in the field."

Each division, as at 3M, has its own product development group. It goes further than that, though. One general manager said. "We're supposed to be centralizing software. But each of my units squirrels away their own capability. They don't feel right without it. Frankly, I look the other way. I don't feel right either. Similarly, each makes its own chips." ("From scratch?" we asked.) "Yes, each starts with raw silicon . . . I worry about the short production runs, the attendant lack of automation. But I'd rather have the new products, even the duplicates. A big hunk of what we're doing 'ought' to be in some other divisions."

The message from the excellent companies we reviewed was invariably the same. Small, independent new venture teams at 3M (by the hundred); small divisions at J&J (over 150 in a $5 billion firm); ninety PCCs at TI; the product champion–led teams at IBM; "bootlegging" teams at GE; small, ever-shifting segments at Digital; new boutiques monthly at Bloomingdale's. That, in a nutshell, is the meaning of chunks. Small *is* beautiful.

Internal Competition. There are fundamentally two ways to sort things out in organizations. The first is "by the rules," or by algorithm, which the rationalists would have us do. It's in the nature of bureaucracy, which is defined as rule-driven behavior, to proceed this way. Thus, we find a 223-committee structure involved in new-product sign-off. At the other end of the spectrum, the "market" is brought inside. The organization becomes driven by internal markets and internal competition. Markets exist for people looking to be assigned to project teams, as at 3M, Fluor, TI, and Bechtel. Direct project competition occurs, as in IBM's "performance shootouts." Bootlegging is not only overlooked but surreptitiously supported, as at GE or IBM. Brands compete at Procter & Gamble. Purposeful overlapping and duplication among divisions and product lines is induced at P&G, Digital, HP, 3M, J&J, and Wang.

An important finding from the excellent companies is the degree to which formal, rational sorting devices are bypassed. For example, at 3M divisions and even groups purposefully compete with one another. Within any group, divisions' charters overlap. ("We would rather have the second product in the market come from another of our divisions.") Managers are specifically rewarded for taking on new-product development activities from *outside* their own division or group.

The idea is an old one. At GM, Alfred Sloan took a crazy-quilt conglomerate of little auto companies and put them into GM's divisionalized structure. He intentionally designed across-the-board overlap: Pontiac with Buick on one end, Pontiac with Chevrolet on the other, and so on. Over the years, GM moved away from the Sloan principles and became more monolithic. One of Roger Smith's stated priorities as recently appointed chairman is to restore the old spirit of competition; he plans to give "free rein to carve out divisional images."

There can be even more intense competition among managers below the division level. At Bloomingdale's, merchandising vice presidents, buyers, and fashion coordinators engage in an unending tussle for scarce floor space. The company reorganizes regularly as both winners and losers emerge.

The prime example of internal competition is brand management at P&G. The company began the formal policy of brand manager competition in 1931, encouraging by official policy "a free for all among brands with no holds barred." Management decided, even then, that internal competition was "the only way to keep from becoming too clumsy." Brand managers today are not given information (other than what is publicly available) about what's going on with other P&G brands. They are encouraged to compete. There is even a special language to describe their competition: "counterpartism," "creative conflict," "the abrasion of ideas." P&G violates the rational rules. One of us made the observation to an ex-P&Ger that brand managers would almost rather cannibalize a fellow brand manager's product than beat the competition. He concurred, adding: "I remember I was a quality control manager when Crest was certified by the American Dental Association a few years back. The next week I ran into a brand manager from one of our other toothpastes. He said, only half kidding, 'Can't you put some bugs in that stuff?'" A large share of new P&G products is likely attributable to the intensity of brand managers' desires to be judged winners. Each year's brand managers become a "class," and competition among classes is fierce.

IBM is the acknowledged master in fostering competition among would-be product ideas. The company formally encourages bootlegging and multiple approaches to the same problem. Then, at some point, it conducts performance "shootouts" among the competing groups—*real* performance comparisons between working hardware and software (not the more typical "competition" among paper plans).

HP has a competitive routine: "Sell it to the sales force." The sales force does not have to accept a product developed by a division unless it wants it. The company cites numerous instances in which several million dollars of development funds were spent by a division, at which point the sales force said, "No, thanks." TI has a similar routine. TI's sales force is also usually separated from the marketing-oriented PCC. TI puts the competitive pressure on by forcing the marketers and product engineers to go straight to the customer, with a car and sales kit, to make the first sales presentations of new prototypes. It's a trial by fire.

A variation on the theme is Digital's willingness to let its segment managers and salesmen push overlapping products onto the product list. Digital lives by an intense user orientation. Therefore, it errs toward tailoring new products to user needs. It does not demand discreteness of products. A *Fortune* analyst notes: "Digital's idiosyncratic growth strategy imposes some penalties. For one thing, many of the 10,000 items on the price list overlap. In certain applications, either of two Digital systems can be used to achieve more or less the same result." So Digital, like P&G, buys into the price of duplication, which is measurable, and assumes (with awesomely regular success) that disproportionate benefits will follow on the revenue line.

Internal competition as a substitute for formal, rule- and committee-driven behavior permeates the excellent companies. It entails high costs of duplication—cannibalization, overlapping products, overlapping divisions, multiple development projects, lost development dollars when the sales force won't buy a marketer's fancy. Yet the benefits, though less measurable, are manifold, especially in terms of commitment, innovation, and a focus on the revenue line.

Intense Communication. A senior HP manager said, "We're really not sure exactly how the innovative process works. But there's one thing we do know: the easy communications, the absence of barriers to talking to one another are essential. Whatever we do, whatever structure we adopt, whatever systems we try, that's the cornerstone—we won't do anything to jeopardize it."

In the excellent companies, there are five attributes of communication systems that seem to foster innovation:

1. *Communication systems are informal.* At 3M there are endless meetings, though few are scheduled. Most are characterized by people casually gathering together—from different disciplines—to talk about problems. The campus-like setting at St. Paul helps, as does the shirtsleeves atmosphere, the no-nonsense midwestern engineering backgrounds, the inbred nature of the organization that ensures that people get to know one another over time. It adds up to the right people being in touch with one another very regularly.

At McDonald's the team at the top lives together informally, setting a tone that pervades the business. At Digital, chief executive Ken Olsen "meets regularly with an engineering committee [consisting of] about twenty engineers from all levels of [Digital]. Olsen sets the agenda and periodically disbands and reconstitutes the committee to maintain a fresh flow of ideas. He sees his role as that of a catalyst or 'devil's advocate.'" One researcher, Ed Schon, states the importance of this sort of interaction in summarizing a major study of the championing process: "Proponents of successful ideas work primarily through the *informal* rather than the formal organization." A championing system at the heart of the organization means a de facto informal culture.

2. *Communication intensity is extraordinary.* Two companies known for their no-holds-barred communications in characteristically uncommunicative industries are Exxon and Citibank. We've had the opportunity to observe senior managers in action at both companies. The difference between their behavior and that of their competitors is nothing short of astonishing. They make a presentation, and then the screaming and shouting begin. The questions are unabashed; the flow is free; everyone is involved. Nobody hesitates to cut off the chairman, the president, a board member.

And how that contrasts with the behavior of most companies we encounter! Senior people, who have sometimes worked together for twenty years or more, won't attend gatherings unless there are formal agendas. They can't seem to do anything other than watch presentations and then politely comment on the contents. At the extreme, people whose offices are on the same floor communicate only in writing. Such behavior contrasts vividly with Cat's daily "no agenda no minutes" meeting among the top ten; Fluor, and Delta's daily "coffee klatch" of the top ten to fifteen; and McDonald's daily informal get-together among the top bunch.

Intel executives call the process "decision making by peers," an open, confrontation-oriented management style in which people go after issues bluntly, straightforwardly. The main reason people need not hide is that they talk all the time. A meeting is not a rare, formal—and thus political—event.

3. *Communication is given physical supports.* A senior IBMer recently shifted jobs, taking on an important research assignment in another high technology company. He walked into an executive's office several weeks after arriving, closed the door, and said, "I've got a problem." The executive blanched; the fellow was critical of his plans. "I don't understand why you don't have blackboards around here," said the ex-IBMer. "How do people talk to each other and exchange ideas without blackboards everywhere?" His point

was well taken. Tom Watson, Sr., started the thrust at IBM with his ubiquitous use of butcher paper on a stand. Physical trappings such as these help spur the intense, informal communication that underpins regular innovation.

The president of a company on our list recounted what he allowed was an important recent activity: "I got rid of the little four-person round tables in the company dining room, replaced them with army mess tables—long, rectangular ones. It's important. At a little round table, four people who already know each other will sit down and eat lunch with each other day in and day out. With long mess tables, strangers come in contact. Some scientist gets talking to some marketer or some manufacturer from some other division. It's a probability game. Every little bit enhances the odds of important idea exchange."

Intel's new buildings in Silicon Valley were designed to have an excess of little conference rooms. Management wants people to eat lunch there, do problem solving there. The rooms are filled with blackboards. (Perhaps we should call this whole set of findings "the blackboard factor.")

MIT's Thomas Allen has been studying physical configurations for years. His results, from research and engineering settings, are striking. If people are more than 10 meters apart, the probability of communicating at least once a week is only to about 8 or 9 percent (versus 25 percent at 5 meters). The figure on the opposite page illustrates the truly dramatic process.

At the most macro level there are disproportionately large numbers of "campuses" among the excellent companies. It's no coincidence, we suspect, that so few of our best performers come from New York City, Chicago, and Los Angeles. Instead, it's Deere's complex in Moline, Caterpillar's Peoria facility, the St. Paul campus of 3M, the P&G setting in Cincinnati, Dana's Toledo center, Dow's headquarters in Midlands, Michigan, HP's beehive in Palo Alto, TI's major Dallas complex, or Kodak's "Kodak Park" in Rochester. In most of these companies, comparatively speaking, many of the important disciplines are gathered together in a single noncosmopolitan setting.

EFFECT OF LOCATION ON COMMUNICATION
R & D and engineering labs

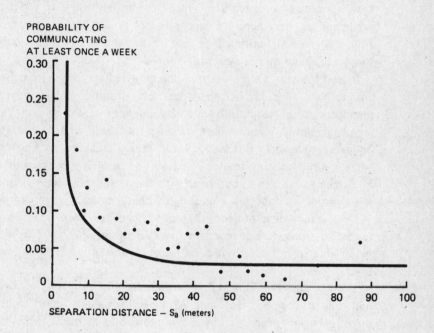

PROBABILITY OF
COMMUNICATING
AT LEAST ONCE A WEEK

SEPARATION DISTANCE — S_a (meters)

4. *Forcing devices.* These are still another aspect of the communication system that spawns innovation, programs that virtually institutionalize innovation. IBM's "Fellow" program is the classic. The IBM Fellows are a manifestation of Watson, Sr.'s, desire to foster "wild ducks" (Watson got the metaphor from Ibsen). There are about forty-five of them, heralded as "dreamers, heretics, gadflies, mavericks, and geniuses" in a recent *Newsweek* ad. "There are less of us than there are corporate vice presidents," said one. A Fellow is given virtually a free rein for five years. His role is quite simple: to shake up the system.

Indeed they do. One of us met one on a red-eye flight from San Jose to New York. He had just spent several million dollars buying microprocessors—essentially out of catalogues—from Silicon Valley companies. "We must have six different labs [at IBM] working on microprocessors. But nobody has really bothered to find out what is already being made. I just sent some of my people out to buy some so we can experiment with them, and play with them." It's amazing, in fact, what one highly charged, crazy man can do. We evaluated some projects that our friend has been involved in (and had someone else confirm our evaluation); he has played a major role in no less than a half-dozen substantial IBM innovations.

The tale goes on. This Fellow has hundreds of people at his beck and call, between San Jose and Armonk. They don't report to him directly, but are available to work on projects as he needs them. His training is as a particle physicist. His favorite activity, time with his customers.

IBM is still a conservative company, though not everyone wears a white shirt any more. Yet this IBM Fellow wears a leather jacket, beads, and a gold chain; he owns two wineries. It must mean that IBM loves him—a lot.

The Individual Contributor program at TI and the New Business Ventures Division at 3M are analogous forcing devices. We have unearthed other examples, as well. Harris and United Technologies give numerous awards for excellence in interdivisional technology transfer. Bechtel urges that every project manager spend fully 20 percent of his time experimenting with new technologies. GE came up with a "toy shop" (a facility in which insiders could view/rent robots) to spur their move into the "factory of the future." Datapoint has installed "Technology Centers" for the same purpose; they are places where people from disparate disciplines are to get together in the name of innovation. All of the above are straightforward efforts to force innovation into the organization.

5. *The intense, informal communication system acts as a remarkably tight control system,* even as it spawns rather than constrains innovation. 3M is a leading example: "Of course, we are under control. No team can spend more than a few thousand dollars without a whole bunch of people looking over their shoulders, not kicking them around, but being genuinely interested in how it's going." We believe that similar "controls" throughout the excellent companies are the truly tight ones. You can't spend much time at one of these companies without lots of people checking up *informally* to see how things are going. In some other companies we know, where controls are more "rigid and formal," you can spend $5 million without bending the first piece of tin and no one will know—as long as you fill out the forms correctly and on time.

TOLERATING FAILURE

A special attribute of the success-oriented, positive, and innovating environment is a substantial tolerance for failure. James Burke, J&J's CEO, says one of J&J's tenets is that "you've got to be willing to fail." He adds that General Johnson, J&J's founder, said to him, "If I wasn't making mistakes, I wasn't making decisions." Emerson's Charles Knight argues: "You need the ability to fail. You cannot innovate unless you are willing to accept mistakes." Tolerance for failure is a very specific part of the excellent company culture—and that lesson comes directly from the top. Champions have to make lots of tries and consequently suffer some failures or the organization won't learn.

One vital observation about failure: it's a lot less punishing with regular dialogue. The big failures, the ones that really leave scars, are usually the ones in which a project was allowed to go on for years without serious guidance. Such eventualities rarely occur in the no-holds-barred communication environment at the excellent companies. The exchange is frank and honest. You can't hide the really bad news, and you don't want or need to.

So the champion's supports are many. The specific devices unearthed number in the hundreds; the evidence presented barely scratches the surface of our data bank. None is a panacea. Each is merely illustrative. The skein of interlocked—and everchanging— supports per se is the message.

Specifically, champions don't automatically emerge. They emerge because history and numerous supports encourage them to, nurture them through trying times, celebrate their successes, and nurse them through occasional failures. But given the supports, the would-be champion population turns out to be enormous, certainly not limited to a handful of creative marvels.

The best imaginable reinforcement for all the major points we've made in this chapter—champions, systems of championing, numbers of experiments, numerous and interlocking supports—can be found in St. Paul, Minnesota. There, 3M has put together an enviable record of financial performance to be sure, but even more so in its constant outpouring of new products. Moreover, the 3M record was not an easy one to put together. It's not the beneficiary of a naturally growing industry or exotic technology; it participates in at least as many slow-growth businesses as fast-growth businesses.

3M—A MAJOR CASE IN POINT

Our study was primarily of giants—the huge corporations, which seldom seem as innovative as they "ought" to be. 3M qualifies as a giant: fifty-first on the *Fortune* 500 list, sales of $6.1 billion in 1980. But 3M *has* innovated: more than 50,000 products in total, well over 100 major new-product offerings each year, 40 plus divisions, with new ones being formed every year. And it *has* been successful. A tidy after-tax profit of $678 million on that just over $6 billion in sales, which puts it fifth in return on sales among the majors (the *Fortune* 100) behind only Sohio, Kodak, IBM, and American Home Products.

3M is in a lot of businesses. The largest, about 17 percent of sales, is tape and allied products, including Scotch Tape. Others are graphic systems, abrasives, adhesives, building materials, chemicals, protective products, photographic products, printing products, static control, recording materials, electrical products, and health care products. But despite the diversity, a common theme prevails at 3M. The company is dominated by chemical engineers who do most of their wizardry with coating and bonding technology. Sticking to that central discipline doesn't mean just mundane product-line extensions. Among the new products within the last two years, *Fortune* notes, are "a suntan lotion that won't wash off when the wearer goes for a swim; a stapler that a surgeon can use to close incisions quickly with metal staples; a film for offset printing that requires no costly silver; and a potion that makes the grass grow slower."

Peter Drucker observes, "Whenever anything is being accomplished, it is being done, I have learned, by a monomaniac with a mission," and 3M fosters the notion that commitment is the sine qua non of good product development. *Fortune* comments on one dimension of that commitment: "What keeps them satisfied in St. Paul is the knowledge that anyone who invents a new product, or promotes it when others lose faith, or figures out how to mass-produce it economically has a chance to manage that product as though it were his or her own business and to do so with a minimum of interference from above."

A part of the champion's support system that we earlier observed as so critical is a protector or buffer of some sort. At 3M one of the protectors is the *executive champion*. Invariably at that company, owing to its history of innovation, the executive champion is an ex-product champion himself, who behaved "irrationally," got shot at, was committed to something, and probably hung in there for ten or more years on some pet project of his own. But now, as the executive champion, he is there to protect the youngsters from premature intrusions from the corporate staff and to push them out of the nest when the time is right. As is so often the case, 3M has a homily or two to describe the executive championing process—for example, "The captain bites his tongue until it bleeds." It's a naval expression and it refers to a junior officer bringing a big ship alongside the dock for the first time. At 3M, it refers to the agonizing process of delegating to the youngsters the all-important activity of nurturing new products. The executive champion at 3M is not a "boss." He is a coach, a mentor. He is paid for his patience and his skill in developing other champions; he is James March's builder of snow fences.

The fundamental unit of support for the champion at 3M is the *new venture team*. It's a task force with some very special characteristics. The three most important: full-time indefinite assignment from various disciplines; volunteers; and staying power.

After a venture team is formed at 3M, it quickly comes to have full-time members from at least the technical area, manufacturing, marketing, sales, and perhaps finance. The team gets full-time members whether it needs them initially or not. The company knows this ritual is apt to be duplicative, especially early on when, say, only a third of a manufacturing person is needed. But they seem willing to pay the price of duplication to get commitment. And only full-time assignment, the sensible 3M argument goes, leads to zealous commitment.

Another marked spur to commitment is that all team members are volunteers. Says a 3M executive, "The team members are recruited, not assigned. There is a very big difference. If I am the marketing person assigned to evaluate the technical guy's idea, in most companies with the usual incentives I can get myself off the hook by saying the idea is poor, by pointing out all the deficiencies. . . . That just doesn't happen if I'm a volunteer team member."

Finally, 3M supports venture team autonomy and staying power. It insists that the team stay together from early in the initiation phase to the eventual rollout. "They say," notes MIT's Edward Roberts, who has studied 3M for twenty years, "we commit to you as a group. You will move forward with the product into the market and benefit from its growth, so long as you meet our conventional corporate measures and standards of performance. In case you fail, we will give you a back-up commitment to job security at the level of the job you left before you entered this venture." (The latter statement depicts another part of the support system: backing for good tries even if they fail.)

The reward system supports both the team and the individual. Everyone gets promoted as a group as their project moves along from hurdle to hurdle. The champion benefits as the group prospers, and vice versa. Here is Roberts again on the subject of the career progress of someone who is a part of a successful venture team:

The individual involved in a new venture will have automatic changes in his employment and compensation categories as a function of sales growth of his product. He starts out, for instance, as a first-line engineer at the top or bottom of the salary range for that job. As his product enters the market he becomes a "product engineer." When annual sales volume hits $1 million, it will automatically become a full-fledged product, and his job title changes. His salary range changes, too, because he now has something selling at $1 million a year. When a product hits the $5 million mark, he passes the next threshold. He is now a "product line engineering manager." If the product reaches $20 million, it suddenly becomes an independent product department, and if he is the key technical person associated with it, he now becomes "manager of engineering or R&D" for that department.

If you want to understand the culture that encourages entrepreneurial activities at 3M, as good a starting point as any is its *value system,* in particular, its "eleventh commandment." It is: "Thou shalt not kill a new product idea." The company may slow it down. Or it may not commit a venture team. But it doesn't shoot its pioneers. As one 3M observer notes, the eleventh commandment is at odds with most activities in large corporations. Moreover, he adds, "If you want to stop a project aimed at developing a new product, the burden of proof is on the one who wants to stop the project, not the one who proposes the project. When you switch the burden from proving that the idea is good to the burden of proving that the idea is not good, you do an awful lot for changing the environment within the company with respect to the sponsorship of entrepreneurial people."

In order to reinforce the shared values clustered around autonomy, innovation, individual initiative, and entrepreneurship, the company's leadership celebrates its heroes—past and present. In our research, one of us sat down with a 3M executive and discussed the last few chairmen and key executives. Virtually without exception, each had a well-publicized championing success. Thus the whole of the top management team, and many of their predecessors, act as role models for the young in the organization. The would-be champion gains encouragement from the panoply of heroes' tales: don't kill ideas; scrounge; failure is OK; years and years are expected to pass before a raw idea makes it in the marketplace; and so on. For instance the tales of the legendary Richard Drew and his cohort John Borden are instructive to the young. Chairman Lewis Lehr relates it: "The salesmen who visited the auto plants noticed that workers painting new two-toned cars were having trouble keeping the colors from running together. Richard G. Drew, a young 3M lab technician, came up with the answer: masking tape, the company's first tape. In 1930, six years after Du Pont introduced cellophane, Drew figured out how to put adhesive on it, and Scotch Tape was born, initially for industrial packaging. It didn't really begin to roll until another imaginative 3M hero, John Borden, a sales manager, created a dispenser with a built-in blade."

This is a typical and surprisingly important vignette, for several reasons. First, it reinforces the close interaction between the company and customer. Second, it shows that the technician doesn't have to be the one who invents. Third, it demonstrates that 3M doesn't limit projects on the basis of potential market size, exactly because the first use (e.g., the first incarnation of Scotch Tape was as a narrow-use industrial fastener) is so often unrelated to eventual product potential. Serious students of innovation note this phenomenon time and time again, with virtually every kind of new product.

When champions win at 3M, they're feted in style. Says Lehr,

"Fifteen to twenty or more times a year some new and promising project reaches a level of a million dollars in profitable sales. You may think that this does not get much attention . . . but it does. Lights flash, bells ring, and video cameras are called out to recognize the entrepreneurial team that is responsible for this achievement." Thus does the company encourage the twenty-eight-year-old engineer with bright ideas to step out and take risks.

3M's value system is also specific in indicating that virtually *any* idea is okay. "Because of 3M's diversity the conviction spreads easily that someone in 3M will be able to use almost anything," a commentator notes. The venerable story illustrating the point is of a failed ribbon material that became a failed plastic cup for brassieres that became *the* standard U.S. worker safety mask after the advent of the Occupational Safety and Health Administration (OSHA). And although the company does stick close to its coating and bonding technological base, it doesn't put any restrictions on the kinds of products it will accept. Roberts notes: "If the product idea can meet financial measures of growth, profitability, and the like, 3M is happy to have it whether or not it's in their dominant field of business." A different point of the same sort surfaced from another 3M executive: "We don't like the cash cow idea. It's the people with success traditions in successful divisions who best realize the potential of continuous innovation." 3M understands that very human truth that success breeds success.

And failure is supported. Legend once more shows the way. Chairman Lehr preaches:

We got into the business of making roofing granules for asphalt shingles because one worker persisted in trying to find a way to use reject sandpaper minerals. He was actually fired [apparently sometimes champions get nailed, even at 3M] because of the time and effort he spent on this. But he kept coming to work anyway. Our Roofing Granules Division today earns substantial revenue. The man responsible retired ten years ago as vice-president of the division. . . . Shortly after World War II, we had a program to develop a bacterial skin barrier, called a surgical drape, for surgeons to use during surgery. The program was twice killed by senior management.* But continued persistence ultimately produced a successful drape and led the way toward our $400 million-a-year health care business today. . . . We keep these stories alive and often repeat them so that any employee with an entrepreneurial spirit who feels discouraged, frustrated, and ineffective in a large organization knows that he or she is not the first one to face considerable odds. . . . The freedom to persist, however, implies the freedom to do things wrong and to fail.

Those who stayed with it were celebrated. Another executive comments: "We don't kill ideas, but we *do* deflect them. We bet on people." And he adds, "You invariably have to kill a program at least once before it succeeds. That's how you get down to the fanatics, those who are really emotionally committed to finding a way— any way—to make it work."

What does it all mean? Among other things, it means living with (managing) a paradox: persistent support for a possible good idea, but not foolish overspending because 3M, above all, is a very pragmatic company. It typically works this way: The champion, as his idea moves out of the very conceptual stage and into prototyping, starts to gather a team about him. It grows to, say, five or six people. Then, suppose (as is statistically the likely case) the program hits a snag. 3M will probably cut it back quickly, knock some people off the team. But as the mythology suggests, the champion—if he is committed—is encouraged to persist, by himself or perhaps with one co-worker, at, say, a 30 percent or so level of effort. In most cases, 3M has observed that the history of any product is a *decade* or more long before the market is really ready. (A decade sounds like a long time, but formal study after formal study reveals that the average space between idea and commercial deployment in virtually any field, high or low technology, is ten to twenty years.) So the champion survives the ups and downs. Eventually, often, the market does become ripe. His team rebuilds.

"We have a belief that we have the capability of solving practical

---

* The ribbon-to-brassiere-to face mask champion was likewise told to knock it off. He ended up doing most of the product development work on the case—at *home.*

problems," says a 3M executive, and that's what 3M is: a company of practical problem solvers, be they salesmen or technical champions. It started that way. One analyst observes: "The obsession with invention dates from the company's origin. Several local investors bought a mine they thought contained valuable corundum, a very hard mineral used in high-grade abrasives. It turned out to be low grade. The investors concluded that the only way to survive was to come up with offshoots that had high value added." Says Lehr, "The salesmen would go from smokestack to smokestack knocking on doors. But they didn't stop at the purchasing agent's office. They went into the back shop to talk to the boys and see what was needed that nobody was making." The salesmen became problem solvers; and the salesman, with his technical buddy in tow, is still the keystone of the 3M strategy today.

3M is the first to recognize innovation as a *numbers game*. "Our approach is to make a little, sell a little, make a little bit more," says Robert M. Adams, vice president of R&D. One of his colleagues talks about "big ends from small beginnings. . . . Spend just enough money to get what's needed next to incrementally reduce ignorance. . . . Lots of small tests in a short interval. . . . Development is a series of small excursions. . . . The odds on any one idea making it through to commercial fruition are approximately zero. . . . There is no limit on raw ideas." So the champions are all over the place experimenting, spending a little. Mostly, they fail. Yet some march through hurdle after hurdle; and a few go all the way.

3M provides funds for people who want to put together a group of any sort, from basketweaving (literally) to solid state physics or micro-electronics. Moreover, the physical "campus" in St. Paul is a hive of pilot testing facilities. The ability to get an idea turned into tin and into a prototype quickly is remarkable. The users, too, are heavily involved in the product development process from its inception through rollout.

In early interviews at 3M, we heard that the average length of a new-product plan was about five pages and were amazed at such brevity. One of us commented on that finding in a speech. A 3M vice president was a speaker, too. He got up and, though generally

supportive of our 3M analysis, said: "You're all wet on that one."
We waited for the other shoe to drop: Did 3M have 200-page new-
product proposals like most of the companies we've worked with?
He went on: "We consider a coherent sentence to be an acceptable
first draft for a new-product plan."

It all works—champions, venture teams, informal communications, voluntary assignment of team members, support for failure,
and the like—because of the incessant focus on keeping the bureaucracy limited. The same vice president added: "We don't constrain
ourselves with plans at the beginning when ignorance is highest.
Sure we plan. We put together meticulous sales implementation
plans. But that's after we know something. At the very front end,
why should we spend time writing a 250-page plan that tries to
drive out ignorance before having first done some simple tests on
customer premises or in a pilot facility somewhere?"

In a similar vein, 3M eschews the idea of a "minimum size" for a
product. "Our experience," says one executive, "tells us that prior
to entry into the market we don't know how to properly anticipate
the sales growth of a new product. Consequently, we tend to make
market forecasts after we've entered the market, not before." And
the head of the New Business Ventures Division stated: "An
NBVD product is *never* justified on the analytic case; it must be
based on belief."

Looked at one way, *organization structure* at 3M isn't important. Roberts observes, "The 3M structure, if you just look at it on
paper, doesn't seem to have anything that is terribly unique." And
in even stronger language, a 3M executive put it, "Structural form
is irrelevant to us."

But there are a number of traits, more or less structural, that are
essential. First, despite a common set of technical disciplines that
might lead others to a functional or matrix organization, 3M remains a radically decentralized business. It has forty or so divisions.
Moreover, the name of the game is creation of new divisions; the
forty is up from about twenty-five just a decade ago. Spinning
things off rather than seeking higher sales volume for one's division
is the time-honored (albeit unconventional) path to success.

That sort of flexibility goes much further, especially in relation to starting up. At 3M, suppose someone working in the product development group in a division comes up with an idea. He first does the normal thing: he goes to his boss to seek funding. Suppose his boss turns him down. Then the 3M magic starts. He goes to another division within his group. If he's turned down again, he goes next to another division within his group. He may be in the adhesives group, but it's not unusual for him to wander over to office products. Now if that group or some other doesn't have time for him, he goes to the court of last resort: the NBVD. That's where the really far-out stuff ends up.

How does 3M make an approach like this work? Simple: managers are given every incentive to do so. The fellow heading any group gets rewarded in part on the dollar amount of venture activity that he's funded from outside his group. The same rule is in force among division heads. Straightforward incentives are there pushing you to look any place to sell an idea, and, if you're a buyer, to look any place to buy one. Concomitantly, the organization is flexible in shifting its people around. After a fellow in Group A sells an idea to a division manager in Group B, say, he moves on over.

There are some associated rules. For example, each division has an ironclad requirement that at least 25 percent of its sales must be derived from products that did not exist five years ago. It is truly remarkable, per conventional theory, that the target is laid on *each* of the over forty divisions (whether in high- or low-growth business).* Such targets in other companies are more commonly applied at the corporate level or at the group level; commitment suffers where it is needed most, at the division where something can be done about it. At 3M where the goal is always demanded at the divisional level, forty separate general managers, not five or ten, are out scrounging for new products.

But the most important notion, as we've said time and again, is that there aren't any one or two things that make it all work. Sure,

---

*This is a P&G trick as well. One former brand manager notes: "The first thing they tell you is, 'Forget product life cycles and cash cows! One of the soaps has been reformulated over eighty times and is thriving."

the champion, the executive champion, and the venture team are at the heart of the process. But they succeed, when they do succeed, only because: heroes abound; the value system focuses on scrounging; it's okay to fail; there's an orientation toward nichemanship and close contact with the customer; there's a well-understood process of taking small, manageable steps; intense, informal communications are the norm; the physical setting provides plenty of sites for experimentation; the organizational structure is not only accommodating but highly supportive of 3M-style innovation; and the absence of overplanning and paperwork is conspicuous, as is the presence of internal competition. That's about a dozen factors. And it's all of them functioning in concert—over a period of decades—that makes innovation work at 3M.

8

Productivity Through People

The Navy, said ex-Chief of Naval Operations Elmo (Bud) Zumwalt, assumes "that everyone below the rank of commander is immature." A friend who runs several plants for General Motors passed on a poem from the auto workers' underground. Its message is poignantly similar:

> Are these men and women
> Workers of the world?
> or is it an overgrown nursery
> with children—goosing, slapping, boys
> giggling, snotty girls?
>
> What is it about that entrance way,
> those gates to the plant? Is it the
> guards, the showing of your badge—the smell?
> is there some invisible eye
> that pierces you through and
> transforms your being? Some aura
> or ether, that brain and spirit washes you
> and commands, "For eight hours
> you shall be different."
> What is it that instantaneously makes
> a child out of a man?
> Moments before he was a father, a husband,
> an owner of property,
> a voter, a lover, an adult.

When he spoke at least some listened.
Salesmen courted his favor.
Insurance men appealed to his family responsibility
and by chance the church sought his help. . . .

But that was before he shuffled past the guard,
climbed the steps,
hung up his coat and
took his place along the line.

The man who gave us this said there was but one key to a people orientation: trust. Some will abuse it. "Three to eight percent, he says, with a smile at the precision of his estimate. Nonbelievers will give you "an infinite number of reasons why workers can't be trusted. Most organizations are governed by rules that assume the *average* worker is an incompetent ne'er-do-well, just itching to screw up." He gives a symbolic illustration: "Ever go to parks? Most are peppered with signs that say, 'Stay off the grass,' 'No parking here,' 'No this,' 'No that.' A few say, 'Campers welcome,' or, 'Picnic tables for your convenience.' One tells you that you *shouldn't*. The other says that you *should,* urges you to join in, take advantage of the facilities." Such a difference in assumptions is monumental in its impact on people, he argues persuasively.

Zumwalt revolutionized the Navy's practices in just a few short years at the helm. It all stemmed from his simple belief that people will respond well to being treated as grownups. He traces his beliefs back to an early command assignment:

What I tried hardest to do was ensure that every officer and man on the ship not only knew what we were about, not only why we were doing each tactical evolution, however onerous, but also managed to understand enough about how it all fitted together that he could begin to experience some of the fun and challenge that those of us in the top slots were having. Our techniques were not unusual. We made frequent announcements over the loudspeaker about the specific event that was going on. At the beginning and the end of the day, I discussed with the officers who, in turn, discussed with their men what was about to happen and what had just happened, what the competition was doing and what we should do to meet it.

We published written notes in the plan of the day that would give the crew some of the color or human interest of what the ship was doing. I had bull sessions in the chief petty officers' quarters, where I often stopped for a cup of coffee. More important than any of these details, of course, was the basic effort to communicate a sense of excitement, fun and zest in all that we were doing.

Zumwalt adds that, within a short eighteen months, practices like this vaulted his ship from being last to first in efficiency within his squadron. "I knew from experience," he said, "the impact of treating sailors like the grown men they were." Tandem's chairman James Treybig sings the same tune: "We assume people are adults." Our Tokyo colleague Ken Ohmae asserts: "Japanese management keeps telling the workers that those at the frontier [first line] know the business best, and that innovation and improvement *must* come from the *genba* (where the action is)." Peter Smith, a recent Wharton MBA grad who eschewed the analyst route and became a General Signal factory manager, agrees: "People will flood you with ideas if you let them."

A work experience related by an MBA student underscores these points (including the typically unhappy ending):

I was operations manager for a major trucking company's San Francisco facility. This terminal was not the leader within the district in any category except unprofitability. I expressed my concerns to some of the Teamsters. They responded by saying that they loved being truckers and felt competent in their roles, but no supervisor had *ever* asked them to help solve the terminal's routing problems or had made them feel as though they were crucial to the operation. My first move with the drivers was to ensure that when they arrived for work in the morning, their tractors were fueled, warmed up, and washed up—ready to go. I hoped the action would impart a sense of urgency to their job. Second, I gave each of them some company caps and brochures to distribute among the customers as they saw fit. (This was strictly forbidden; only salesmen could do this. I had to steal the caps from a salesman's car one morning.)

Most important, supervisors had traditionally routed all the local freight (usually unsuccessfully); I instructed them to leave every third or fourth freight bill unrouted, so that when they were asked for routing instructions by the dockman, they could gracefully ask for suggestions. I kept most of these ideas secret from my bosses and the union hierarchy. To my surprise, the operation became profitable. I posted the financial figures on the *union* bulletin board (again, strictly against the rules) and never received a complaint. It even got to the point where the salesmen realized that the drivers were soliciting more new customers than they were, so several decided to ride with the drivers to learn their "secrets."

The profitability lasted for several periods, until my boss saw what was happening and became nervous of the leeway given the Teamsters. About that time, the company instituted a control system that required each Teamster to account for every fifteen minutes of his work day. Profitability disappeared and customer complaints increased. I left for school.

Treat people as adults. Treat them as partners; treat them with dignity; treat them with respect. Treat *them*—not capital spending and automation—as the primary source of productivity gains. These are fundamental lessons from the excellent companies research. In other words, if you want productivity and the financial reward that goes with it, you must treat your workers as your most important asset. In *A Business and Its Beliefs,* Thomas J. Watson, Jr., puts it well: "IBM's philosophy is largely contained in three simple beliefs. I want to begin with what I think is the most important: *our respect for the individual.* This is a simple concept, but in IBM it occupies a major portion of management time. We devote more effort to it than anything else. This belief was bone-deep in my father."

There was hardly a more pervasive theme in the excellent companies than *respect for the individual.* That basic belief and assumption were omnipresent. But like so much else we have talked about, it's not any one thing—one assumption, belief, statement, goal, value, system, or program—that makes the theme come to life. What makes it live at these companies is a plethora of structural devices, systems, styles, and values, all reinforcing one another so that the companies are truly unusual in their ability to achieve extraordinary results through ordinary people. The message goes right back to our early chapter on man and motivation. These companies give people control over their destinies; they make meaning for people. They turn the average Joe and the average Jane into winners. They let, even insist that, people stick out. They accentuate the positive.

Let us make clear one final prefatory point. We are not talking about mollycoddling. We are talking about tough-minded respect for the individual and the willingness to train him, to set reasonable and clear expectations for him, and to grant him practical autonomy to step out and contribute directly to his job.

Genuine people orientation is in marked contrast to the two major alternatives all too often seen in companies: the lip service disaster and the gimmicks disaster.

The lip service disaster is arguably the worse of the two. Almost every management we've been around says that people are important—vital, in fact. But having said that, they then don't pay much attention to their people. In fact, they probably don't even realize their omissions. "People issues take up all my time," is the typical rejoinder. What they often really mean is, "This business would be so easy if it weren't for people."

Only when we look at the excellent companies do we see the contrast. The orientation toward people in these companies often started decades ago—full employment policies in times of recession, extraordinary amounts of training when no training was the norm, everybody on a first-name basis in times much more formal than ours, and so on. Caring runs in the veins of the managers of these institutions. People are why those managers are there, and they know it and live it.

The orientation is bone-deep and embedded in the language itself. At Delta, it's the "Family Feeling." At Hewlett-Packard, it's "the HP Way," and "Management by Wandering Around." At Dana, it's simply the constant use of the word "people"—in annual reports, in top executive speeches, in statements of policy. (Rene McPherson, ex-chairman, is vehement about it. In a casual conversation he brings up a new Ford blockbuster ad campaign. "Damn it," he says. "They talk about 'workers'. Why not 'people'?") Employees are called "crew members" rather than personnel at McDonald's, "hosts" at Disney Productions, and "associates" at J. C. Penney.

Although it may be corny, it's unabashed hoopla and people respond to it. When we first looked at the phenomenon, we thought large doses of hoopla and celebration might be limited to companies like Tupperware, where the president and his senior managers are said to participate for thirty days a year in Jubilees, aimed at feting the success of their top *15,000* salespersons and managers. But we found hoopla going on in the high tech companies as well (e.g., HP's song, "Grab a Grizzly," in celebration of its 3000 series computer). And at Caterpillar we were told of an event to introduce new equipment where huge pieces of earth-moving machinery were dressed in costume.

Perhaps surprisingly, the people orientation also has a *tough* side. The excellent companies are measurement-happy and performance-oriented, but this toughness is borne of mutually high expectations and peer review rather than emanating from table-pounding managers and complicated control systems. The tough side is, in fact, probably tougher than that found in the less excellent and typically more formal systems-driven companies, for nothing is more enticing than the feeling of being needed, which is the magic that produces high expectations. What's more, if it's your peers that have those high expectations of you, then there's all the more incentive to perform well. People like to compare themselves to others, as we noted in Chapter 3, and they also like to perform against standards—if the standard is achievable, and especially if it is one they played a role in setting.

The point, then, is the *completeness* of the people orientation in the excellent companies. In lip service institutions, no matter what they say, almost all of what we have just described is missing. Layoffs are certainly not taken lightly, but we find few stories that parallel IBM's, Delta's, Levi's, or HP's truly unusual efforts to avoid the ups and downs of employment. And the language *is* different. The war stories in the less well performing institutions do not refer nearly so much to the care, handling, and feeding of employees as they do at a Dana, a Digital, or an IBM. The word "manager" in lip service institutions often has come to mean not someone who rolls up his or her sleeves to get the job done right alongside the worker, but someone who hires assistants to do it. These companies never mention peer review. They are secretive and purposely hide information from employees. The message here is clear: the employees supposedly aren't grown up enough to handle the truth. And the hoopla, razzle-dazzle, and constantly changing menu of prizes, awards, and other incentives? Missing as well. Sure, sometimes a new program like MBO, or quality circles, or Scanlon Plan is tried when it becomes faddish. But each is soon rejected or bureaucratized. The failing is often attributed to "the unions," or "lack of employee goodwill." Seldom is it attributed to lack of persistence and true caring on the part of management.

That leads directly to the second problem: the gimmicks trap. The current gimmick is the quality circle. There is absolutely nothing wrong with the idea, as the Japanese have so forcefully reminded us. But quality circles are only the latest in a long line of tools that can either be very helpful, or can simply serve as a smokescreen while management continues to get away with not doing its job of real people involvement. Ten years ago, it was job enlargement. Before that it was the seemingly ubiquitous organization development movement, replete with team building, T-groups, conflict resolution, and managerial grids. The bones of these programs are scattered on America's low-productivity desert. Very little has changed. Consultants and other practitioners sold their programs to lower levels of management, like the training officers, and top management let them go ahead with it, as much to avoid getting their own hands messy as anything else. But these supposed panaceas could not successfully be applied in a wholly bottom-up way—that is to say, applied without intense top management interest. It simply won't work. The implicit changes required are nothing short of earth-shaking. There is no way that such programs will ever take hold without the unstinting support of the whole top management team.

As there is no way that just a few programs will take hold and bring about fundamental change, so also there is no reason to expect any particular technique to have an effective life of more than a few years. Most of the excellent companies *do* have MBO systems, and they *do* have quality circles, and they probably *have* tried team building, and maybe they still use all of these. But they have lots more. We were astounded, as we did our research, by the sheer number of people programs we encountered and the frequency with which they are replenished or refurbished. And these programs are neither lip service nor gimmicky. We found rich systems of monetary incentives; but we expected that. We also discovered an incredible array of nonmonetary incentives and an amazing variety of experimental or newly introduced programs. No one device—even in the best institutions—is likely to be effective indefinitely. The point is to treat the problem as one would the new-product challenge. The pipeline must always be filled with the next score of candidate programs, most of which will turn out to be duds, just as do new-product ideas. If job enrichment doesn't work at the Milwaukee plant, try seven other programs that are working in other plants, or that have worked in other companies.

SUCCESS STORIES

Although most top managements assert that their companies care for their people, the excellent companies are distinguished by the intensity and pervasiveness of this concern. The only way to describe it adequately is through example.

RMI

RMI is a good one to start with. A subsidiary of U.S. Steel and National Distillers, it is an integrated producer of titanium products. For years its performance was substandard. Poor productivity, poor profits. But in the last five years RMI has had a remarkable success, owing almost entirely to its adoption of an intensely people-oriented productivity program.

The program started when "Big Jim" Daniell, a former professional football player, ex-captain of the Cleveland Browns, was made chief executive. The program he installed was described by *The Wall Street Journal* as "pure corn—a mixture of schmaltzy sloganeering, communication, and a smile at every turn." His plants are peppered with notices that say: "If you see a man without a smile, give him one of yours," or: "People rarely succeed at anything unless they enjoy it." All are signed "Big Jim."

The story doesn't get much more complicated than that. The company's logo is a smile-face, which is on the stationery, on the front of the factory, on signs in the factory, and on the workers' hardhats. RMI's headquarters is in Niles, Ohio, which everyone now calls "Smiles, Ohio." Big Jim spends much of his time riding around the factory in a golf cart, waving and joking with his workers, listening to them, and calling them all by their first name—all 2,000 of them. Moreover, he spends a lot of time with his union. The local union president paid him the following compliment: "He calls us into his meetings and lets us know what's going on, which is unheard of in other industries."

What's the result of it all? Well, in the last three years, with hardly a penny of investment spending, he's managed an almost 80 percent productivity gain. And at last report, his average backlog of union grievances had declined from about 300 to about 20. Big Jim, say those of his customers that we've come across (e.g., at Northrop), simply exudes care about his customers and his people.

Hewlett-Packard

In one study, eighteen out of twenty HP executives interviewed spontaneously claimed that the success of their company depends on the company's people-oriented philosophy. It's called "the HP

Way." Here's how founder Bill Hewlett describes it:

I feel that in general terms it is the policies and actions that flow from the belief that men and women want to do a good job, a creative job, and that if they are provided with the proper environment they will do so. It is the tradition of treating every individual with consideration and respect and recognizing personal achievements. This sounds almost trite, but Dave [co-founder Packard] and I honestly believe in this philosophy. . . . The dignity and worth of the individual is a very important part, then, of the HP Way. With this in mind, many years ago we did away with time clocks, and more recently we introduced the flexible work hours program. Again, this is meant to be an expression of trust and confidence in people as well as providing them with an opportunity to adjust their work schedules to their personal lives. . . . Many new HP people as well as visitors often note and comment to us about another HP way—that is, our informality, and our being on a first name basis. I could cite other examples, but the problem is that none by themselves really catches the essence of what the HP Way is all about. You can't describe it in numbers and statistics. In the last analysis it is a spirit, a point of view. There is a feeling that everyone is part of a team, and that team is HP. As I said at the beginning, it is an idea that is based on the individual. It exists because people have seen that it works, and they believe that this feeling makes HP what it is.

The people orientation at HP started early. In the 1940s Hewlett and Packard decided "not to be a hire and fire company." That was a courageous decision in those times, when the electronics business was almost entirely government-supported. Later, HP's collective mettle was to be tested when business was severely down during the 1970 recession. Rather than lay people off, Hewlett, Packard, and everyone else in the organization took a 10 percent cut in pay. Everyone worked 10 percent fewer hours. And HP successfully weathered the recession without having to sacrifice full employment.

The people philosophy at HP not only began early on but is also self-renewing. The corporate objectives were just rewritten and republished for all the employees, including a restatement of corporate philosophy. The very first sentence reads: "The achievements of an organization are the result of the combined efforts of each individual . . . ." And a few sentences later HP reinforces its commitment to innovative people, a philosophy that has been a driving force in the organization's success. "FIRST, there should be highly capable, innovative people throughout the organization . . . SECOND, the organization should have objectives and leadership which generate enthusiasm at all levels. People in important management positions should not only be enthusiastic themselves, they should be selected for their ability to engender enthusiasm among their associates." The introduction to the revised corporate objective statement concludes: "Hewlett-Packard [should not] have a tight, military-type organization, but rather . . . give people the freedom to work toward [overall objectives] in ways they determine best for their own areas of responsibility."

The faith that HP has in its people is conspicuously in evidence in the corporate "open lab stock" policy that a few of our students encountered in the Santa Rosa division. The lab stock area is where the electrical and mechanical components are kept. The open lab stock policy means that not only do the engineers have free access to this equipment, but they are actually encouraged to *take it home for their personal use!* The idea is that whether or not what the engineers are doing with the equipment is directly related to the project they are working on, by fooling around with the equipment at work or at home, they will learn—and so reinforce the company's commitment to innovation. Legend has it that Bill [Hewlett]* visited a plant on a Saturday and found the lab stock area locked. He immediately went down to maintenence, grabbed a bolt cutter, and proceeded to cut the padlock off the lab stock door. He left a note that was found on Monday morning: "Don't ever lock this door again. Thanks, Bill."

The same language pervaded a conversation with a twenty-four-year-old engineer, on the scene for barely more than a year. Commenting on some problems with a new personnel procedure, he said:

---

*All Hewlett or Packard stories, regardless of the teller's age, refer to "Bill" or "Dave."

"I'm not sure Bill and Dave would have done it that way." It's truly remarkable to find the value set stamped in so quickly, and with such clarity. The young man went on to describe HP's dedication to "getting on with it," the need to be involved with successful new-product introductions in order to get ahead, the litany of succeeding by a record of hard accomplishments rather than paper-pushing skills, the ability to talk to anyone, anywhere. He talks of his division's general manager and senior officers as though they were close friends and he were their only employee. He rambles on about MBWA. The discussion drifts to such publicly touted communications devices as the "coffee klatch," where informal problem solving (all hands attending) takes place weekly. The PR hype turns out to be justified.

In short, the most extraordinary trait at HP is uniformity of commitment, the consistency of approach and attitude. Wherever you go in the HP empire, you find people talking product quality, feeling proud of their division's achievements in that area. HP people at all levels show boundless energy and enthusiasm, so much so that many of our colleagues, after a chance encounter with an HP executive, engineer, or line worker, ask: "Is this guy for real?" And then they meet more, and invariably their skepticism, no matter how hard they try to keep it, begins to fade. We ourselves tried to remain sober, not to become fans. But it proved impossible.

Wal-Mart

Wal-Mart, with over 26,000 employees, is now the number four retailer in the United States. During the 1970s, growth took the company from $45 million in sales to $1.6 billion, from 18 stores to 330. Sam Walton, or "Mr. Sam," as he is called in the company, is the driving force behind this success, and Walton, quite simply, cares about his employees. In fact, almost all his managers, at his insistence, wear buttons that say, "We Care About Our People."

Walton learned the people business at J. C. Penney. Like Penney's, his people are referred to as "associates," not employees. And he listens to them. "The key is to get out into the store and listen to

what the associates have to say," he says. "It's terribly important
for everyone to get involved. Our best ideas come from clerks and
stockboys." Walton stories have become legends. According to *The
Wall Street Journal:* "Mr. Walton couldn't sleep a few weeks back.
He got up and bought four dozen donuts at an all night bakery. At
2:30 A.M., he took them to a distribution center and chatted for a
while with workers from the shipping docks. As a result he discovered that two more shower stalls were needed at that location."
Again, the astonishing point is not the story per se: any small business person could relate a host of similar tales. The surprising news
is that a top executive still exhibits such a bone-deep form of concern for his people in a *$2 billion* enterprise.

The message that down-the-line people count is mirrored in every
activity. The executive offices are virtually empty. Headquarters
resemble a warehouse. The reason is that Walton's managers spend
most of their time out in the field in Wal-Mart's eleven state service areas. And what are they doing? "Leading local cheerleading
squads at new store openings, scouting out competing K mart
stores, and conducting soul-searching sessions with the employees."
Walton himself visits every store every year (330 now, remember)
as he has done since 1962.

Everyone at Wal-Mart feels like a winner. The regular management meetings start at 7:30 A.M. on Saturday. The buyer of the
month receives a plaque. There are "honor roll" stores, every *week*.
And every week the "SWAT" team that swoops down to remodel
stores testifies to jobs well done. Mr. Sam stands up and yells,
"Who's number one?" And everyone, of course, yells back "Wal-
Mart!"

So, it's intense rah-rah, and, yes, it's hocum, and—like so many
other situations we see—it's fun. As *The Wall Street Journal* reports: "Mr. Walton seems to have the most fun. Not long ago he
flew his aircraft to Mt. Pleasant, Texas, and parked the plane with
instructions to the co-pilot to meet him 100 or so miles down the
road. He then flagged a Wal-Mart truck and rode the rest of the
way to 'Chat with the driver—it seemed like so much fun.'"

The theme of fun in business runs through a great deal of the

excellent companies research. The leaders and managers like what
they do and they get enthusiastic about it. Or, as Howard Head
said in a recent speech, "It seems to me you have to be personally
associated with what you do. I just love design. If it weren't fun, I
wouldn't do it."

Dana

One of the most impressive success stories in people and productivity is that of the Dana Corporation under the leadership of Rene
McPherson. Dana is a $3 billion corporation, making unexotic
products like brass propeller blades and gearboxes, primarily supporting the unexciting secondary market in the automobile and
trucking industry. If you had looked at Dana as a proposition in
strategic management, you would undoubtedly have labeled it a loser. Yet in the 1970s, this old-fashioned midwestern business became the number two *Fortune* 500 company in total return to investors. In the early 1970s, the sales per employee at Dana were the
same as the all-industry average. By the late 1970s, and without
massive capital spending, Dana's sales per employee had tripled
while the all-industry average had not even doubled, (in Dana's
industry segment, productivity had barely increased), a phenomenal
productivity record for a huge business in an otherwise uninterest-
ing industry. Furthermore, Dana is largely unionized, with the
United Auto Workers (UAW) in most of its plants. But during the
same decade, its grievance rate fell to a tiny fraction of the overall
UAW average.

The key ingredient is productivity through people, pure and simple. As we mentioned earlier, when McPherson took over in 1973,
one of his first acts was to destroy 22½ inches of policy manuals
and substitute a simple one-page statement of philosophy. It reads
in the main:

• Nothing more effectively involves people, sustains creditability or
  generates enthusiasm than face to face communication. It is critical to provide and discuss all organization performance figures
  with all of our people

- We have an obligation to provide training and the opportunity for development to our productive people who want to improve their skills, expand their career opportunities or simply further their general education
- It is essential to provide job security for our people
- Create incentive programs that rely on ideas and suggestions, as well as on hard work, to establish a reward pool.

Says McPherson: "The philosophy comes first. Almost every executive agrees that people are the most important asset. Yet almost none really lives it."

McPherson quickly reduced his corporate staff of 500 to 100 and the number of layers in his organization from eleven to five. His plant managers—about ninety of them—all became "store managers." In a litany repeated at Delta and at Disney, they were made responsible for learning *all* the jobs in the plants. And they were given the autonomy to get the overall job done. Their success led McPherson to say, in a statement that could get someone else kicked out of most board rooms in America, "I am opposed to the idea that less government, fewer regulations, capital formation incentives, and renewed research in development activity are what we need most to improve productivity. My suggestion: let our people get the job done."

At Dana, philosophy does come first; but then it's largely a matter of a voluntary diffusion of ideas. Everyone is responsible for ensuring that productivity increases take place. McPherson suggests the appropriate starting point: "Personal productivity of the top managers is a vital symbol." But nobody is told how to do it. If there is a how, it is a simple belief in the inherent will toward efficiency of the man down at the bottom of the organization. As McPherson points out:

Until we believe that the expert in any particular job is most often the person performing it, we shall forever limit the potential of that person, in terms of both his own contributions to the organization and his own personal development. Consider a manufacturing setting: within their 25-square-foot area, nobody knows more about how to operate a machine, maximize its output, improve its quality, optimize the material flow and keep it operating efficiently than do the machine operators, material handlers, and maintenance people responsible for it. Nobody.

He adds:

We didn't waste time with foolishness. We didn't have procedures, we didn't have lots of staff people. We let everybody do their job on the basis of what they need, what they say they'll do, and what their results are. And we gave them enough time to do it. . . . We had better start admitting that the most important people in an organization are those who actually provide a service or make and add value to products, not those who administer the activity. . . . That is, when I am in your 25 square feet of space, I'd better listen to you!

McPherson's focus is always the same. In casual conversation or formal presentation, he never wavers from his emphasis on people. As one of his former associates at Dana said to us, "I never heard him make a statement that didn't say something about people." McPherson says, "Look at the pictures in the annual reports. Don't worry about the chairman; he always gets his name under the picture—and it's spelled right, too. Look for pictures of people [down-the-line workers]. How many of them are identified by name?"

Like HP, Dana did away with time clocks. "Everybody complained," McPherson says. "'What do we do without time clocks?' I said, 'How do you manage any ten people? If you see them come in late regularly, you talk to them. Why do you need time clocks to know if the people who work for you are coming in late?'" He also reinforces the focus on starting from positive assumptions about people's behavior as he elaborates on the story: "My staff said, 'You can't get rid of the time clocks. The government requires a record of every person's attendance and time worked.' I said, 'Fine. As of now, everyone comes to work on time and leaves on time. That's what the record will say. Where there are big individual exceptions, we will deal with them on a case-by-case basis.'"

McPherson is a bug on face-to-face communication and on discussing *all* the results with *all* of the people. He required that there be a monthly face-to-face meeting between division management and *every* member of the division to discuss directly and specifically all of the detailed corporate individual results. (We see that time and again in the excellent companies. They are obsessed about widely sharing information and preventing secrecy. They willingly trade any marginal loss of competitive information for the added commitment.) McPherson even stressed face-to-face contact in institutional advertisements. He ran ads that, as he says, "made my middle managers *very* nervous at first." One said: "Talk Back to the Boss," another: "Ask Dumb Questions." McPherson deplores management's unwillingness to listen: "I wanted a picture, for a slide presentation, of a worker talking to his foreman. We had fourteen thousand photos in the file, but not one of a supervisor *listening* to a worker."

McPherson spent 40 to 50 percent of his time on the stump, carrying the message directly to his people. He insisted on what he called "Town Meetings," with everybody in attendance. He recalls an experience in Reading, Pennsylvania: "I wanted to talk to all the people. The boss said there is no place to do it. It went on that way for three years. Finally, I said, 'Clean out the shipping department.' Sixteen hundred people showed up. In all my years of travel, I never got one cheap shot question from an employee. Yet my plant manager and division manager, when I insisted they go with me, never wanted to go. . . . Look at these pictures," he adds, pushing the stack over to us. "They are from the meetings. Always machine operators, never managers asking questions. You know why? Managers won't ask questions. They're scared."

Another McPherson obsession is training, continuous self-improvement. McPherson's pride and joy is Dana University. Several thousand Dana employees trooped through Dana U. last year. Classes are practical, but at the same time they reinforce the people philosophy. Many classes are taught by seniors—corporate vice presidents (we found a similar phenomenon at Disney U. and McDonald's Hamburger U.). According to McPherson, there is no more prestigious position for any member of the management than an appointment as a regent of Dana University. The Board of Regents is usually composed of nine division general managers.

Nothing is forced at Dana. The Scanlon Profit Plan, for which Dana has gotten a lot of publicity, is a good example. Much to our surprise, it turns out that the Scanlon Plan is only in seven of the forty Dana divisions. McPherson says: "They go where they work. That's all. No division manager is under pressure to accept one."

The major pressure at Dana—and it's a very real one, as in most of our other excellent companies—is peer pressure. Dana's effort to foster it is capped by Hell Week. Twice a year about a hundred managers get together for five days to swap results and productivity improvement stories. McPherson encouraged the process, because he believes that peer pressure is what makes it all go. He says, "You can always fool the boss. I did. But you can't hide from your peers. They know what's really going on." And, of course, there is free and open communication, bordering on a free for all, during Hell Week. He ran ads that supported this one, too: "We put them through hell."

McPherson's philosophy on job security has been tested severely during the recent hard times in the American auto industry. Much as the company would have liked to avoid it, it had to lay people off. On the other hand, even those actions were accompanied by continued intense communications. Everyone was told what was going on—as it happened. Says McPherson of the practical results, "We had an eighty percent participation in the stock plan in 1979. Then there were nine thousand layoffs. What's our participation rate now, including those laid off? Still eighty percent." Moreover, the 1981 bounce back in results by Dana, going strongly against the tide, is truly phenomenal.

The McPherson philosophy comes down to the value of everyone's contributing ideas, not just keeping up the pace on the line. "The way you stay fresh," stresses McPherson, "is you never stop traveling, you never stop listening. You never stop asking people what they think." Contrast that with the following comment from a General Motors worker, recently laid off after sixteen years in the Pontiac division: "I guess I got laid off because I make poor-quality cars. But in sixteen years, not once was I ever asked for a suggestion as to how to do my job better. Not once."

Delta Airlines

Delta Airlines is one of the handful to go through deregulation in the airline industry with few scars on its unblemished record of strong financial performance. Delta's last strike was in 1942. The last union vote was in 1955. Francis O'Connell of the Transport Workers of America says of Delta: "[They have] a relationship with their employees that is most difficult to break into."

Delta is a people company. It advertises "the Delta Family Feeling," and lives that philosophy. The company promotes from within, pays better than most airlines, and goes to any length to avoid laying workers off in a traditionally cyclic industry.

As many of the excellent companies do to ensure a fit with the culture, Delta begins with a careful and lengthy screening process for all job applicants. *The Wall Street Journal* notes, "Stewardesses, for instance, are culled from thousands of applicants, interviewed twice and then sent to Delta psychologist, Dr. Sidney Janus. 'I try to determine their sense of cooperativeness or sense of team work. At Delta, you don't just join a company, you join an objective.'"

Success at Delta stems from a collection of lots of little things. The open door policy sets the tone. Ex-president Tom Beebe explains: "My rug has to be cleaned once a month. Mechanics, pilots, flight attendants—they all come in to see me. If they really want to tell us something—we'll give them the time. They don't have to go through somebody. The chairman, president, vice president—none of us has a single 'administrative assistant' to screen people out, no intermediaries." Of course, what makes it work is that something *happens* when the open door is used. Delta spends a lot of time and money (entirely inconceivable to those who don't practice such things) checking out the employee's side of the story. Often, the result is a substantial policy change—surrounding, say, pay or accounting procedures. It's all "brought about by the time-honored willingness of employees to use the open door and the time-honored willingness of top management to keep the door open," says one analyst.

Here is one very typical example of the policy at work, as reported by *The Wall Street Journal:*

In February 1979, James Burnett's paycheck came up $38 short. Delta Airlines hadn't paid him enough overtime for the day he came in at 2:00 A.M. to repair an L-1011 engine. When his supervisor wouldn't help, the 41-year-old mechanic wrote Delta's President, David C. Garrett, Jr. He complained that "the pay problem we have experienced is bad and it has caused a lot of good men to go sour on the company." Three days later, Mr. Burnett got his money and an apology from top management. Delta even changed the pay policy, increasing overtime pay for mechanics called in outside normal working hours.

One of the more interesting notions at Delta is that of interchangeability of management parts. The chairman insists, for example, that all his senior vice presidents be trained to step into any job in the company (though not, presumably, flying the planes). Even the senior vice presidents are supposed to know one another's areas well enough to substitute for any other if need be. And, incidentally, it is a tradition for top management to pitch in and help baggage handlers at Christmas time.

Like Dana, Delta management spends an extraordinary amount of time just plain talking to its people. Senior management meets with all employees at least once a year in "open forum," where direct communications take place between the highest and lowest levels of the organization. The amount of management time required for all these communications is staggering and, again, difficult to imagine for those who don't work in this kind of environment. For example, very senior management holds four full days of meetings a year just to talk to flight attendants based in Atlanta. Senior vice presidents typically spend more than a hundred days a year on the road; they are not easy days, but include time down at the flight line at 1:00 or 2:00 A.M., checking out the graveyard shift. Intense communications start at the top. There is a ritual Monday morning staff meeting where all the company's programs, problems, finances are thoroughly reviewed. Afterward, senior vice presidents take their department heads to a late lunch to bring them up to date. And so the news is quickly and regularly passed through the company.

Listening to employees is taken seriously. For instance, a committee of flight attendants chooses uniforms for Delta's 6,000 stewardesses and stewards. "That's important, you have to live in them," said one flight attendant. Mechanics even choose their immediate supervisor.

McDonald's

It seems appropriate that Fred Turner, McDonald's current chairman, started out as a shoe salesman. In such ways the leaders in many people-intensive organizations learned what it was to get the basics right—to meet customers, provide real-time service, and take pride in and responsibility for a mundane job. McDonald's is, above all, better at the basics. Says Turner: "History shows that [the competitors'] management involvement doesn't last. They just don't have the depth of attention to detail."

McDonald's believes that senior managers should be out in the field, paying attention to employees, training, and execution. Says the founder, Ray Kroc, "I believe that *less is more* in the case of corporate management; for its size, McDonald's today is the most unstructured corporation I know, and I don't think you could find a happier, more secure, harder working group of executives anywhere."

McDonald's talks endlessly about the individual's contribution. Kroc argues, "A well-run restaurant is like a winning baseball team, it makes the most of every crew member's talent and takes advantage of every split-second opportunity to speed up service." Kroc focuses on the little things: "I emphasize the importance of details. You must perfect every fundamental of your business if you expect it to perform well." Getting the details right, McDonald's way, requires an astonishing amount of learning and intensity. Says a former employee, "When I first started, they put a little white hat on me that said 'trainee.' They started me right off in the easiest of the jobs—cooking french fries. Then I moved to fries and shakes. So it went, on up to handling the buns and cooking the burgers. We only had one small room where we could take breaks. There was a TV and cassette going on all the time, stressing some aspect or another of the way McDonald's does things. How to cook a better burger, how to keep the fries crisp, the whole bit."

"The book" at McDonald's spells out procedures and details. For instance, "Cooks must turn, never flip, hamburgers." Or, "If they haven't been purchased, Big Macs must be discarded in ten minutes after being cooked and french fries in seven minutes. Cashiers must make eye contact with and smile at every customer." And on it goes.

Despite the rigidity of procedure surrounding many such areas, store managers are encouraged to exercise autonomy and keep things lively. *Fortune* reports that "Debbie Thompson, who started out at McDonald's as a cashier eight years ago, and now, at 24, manages the company-owned store at Elk Grove Village, sometimes livens up the lunchtime rush hour by offering $5 bonuses to the cashiers for taking in the most dollars and handling the most customers. She gives a plaque to the crew member of the month." *
Another employee adds: "We always got paid a dollar for making a record amount of sales for an hour. Also, if you had a three hundred dollar hour [in food sales] you got a dollar. Everyone working in that period got a dollar. On the record days you got two dollars. We were all shooting for the extra bucks. It meant something."

A vital part of the system is Hamburger U. The *New York Times* reports:

The American flag and the McDonald's flag fly high over the expressway running through the backyard of Hamburger University in a suburban Chicago town. Inside, McDonald's franchisers and company managers learn skills to reinforce what the golden arches of 614,000 similar brick buildings in mostly suburban and rural communities have come to symbolize: predictability in atmosphere and taste, or, as McDonald's founder Ray Kroc put it, "the gospel of Quality, Service, Cleanliness and Value." A high school drop

---

* Some consider such rewards trivial. However, a Stanford MBA student remembers winning a similar reward at Jack In The Box. "It may sound silly, but I have carried it around with me for seven years." We have a friend, a salesman, who won a barbecue as part of a sales contest. His barbecue at home was much better than the model he won. Nonetheless, he ripped out the previous one and replaced it with the new one, his prize.

out, Mr. Kroc has donated millions to charity and urges employees to become active in community charities to further McDonald's image, but he refuses to support higher education. In his book [*Grinding It Out*] he writes, "One thing I flatly refused to give money to is the support of any college. I have been wooed by some of the finest institutions in the land but I tell them they will not get a cent from me unless they put in a trade school." . . . Two thousand students "graduated" from the school [Hamburger U.] last year. . . . One lucky student in each course receives a golden chef's hat for making the largest contribution to class discussion. . . . Another walks away with a ceramic abstract model of a hamburger for highest academic honors. . . . McDonald's points to the fact that the American Council on Education recommends college credit of up to six semester hours for Hamburger U. courses taken by those pursuing a degree in two or four year colleges . . . [there are] 18 courses from one or two day seminars to week long sessions on "market evaluation," "management skills," and "area supervision." . . . McDonald's success is based on fast food and friendly service at a low price. Courses deal with McDonald's style and emphasize motivation. . . .

McDonald's also turns to hoopla and razzle-dazzle. As one employee recalls:

One of the guys in our store was an "All-American Hamburger Maker." He was the best hamburger cooker in McDonald's chains across the country. The competition begins in the spring. They have an All-American contest to see who is the best, literally the best hamburger cooker in the country. It means the quickest, but also the most nearly perfect, the top quality, cooking them exactly the way they are supposed to be cooked. To do it really right you get a little thermometer and you stick it on top of the grill. The grill would be shining, absolutely spotless. Then you lay out the burgers just so, six in a row, perfectly in line. You sear them all with the back of a spatula, you salt them at the right moment, put the onions on at the right moment. Then you take them off properly, lay them on the buns. . . . First, you have the in-store competition to find the best hamburger cooker in the store. The guy who won that then goes on to the regional championships. Then they go to the next level. Finally, they go to the All-American contest—I think it

was in Chicago. There was a big trophy involved, and I think there
was money involved, but I don't know how much. The important
thing was that you got to wear an All-American patch on your
shirt.

IBM

From McDonald's we circle all the way to IBM, perhaps one of the
biggest and oldest American companies practicing an intense people orientation. The only issue with IBM is how to start describing
it. With the seventy-year-old open door policy? The senior Mr.
Watson's $1-a-year country club, established for all employees in
the 1920s? The philosophy that starts with "respect for the individual"? Lifetime employment? Insistence upon promotion from within? IBM day-care centers, IBM hotels, IBM running tracks and
tennis courts? *Monthly* opinion surveys by the personnel department? A very high success rate among salesmen? The intense training? IBM's total history is one of intense people orientation. And as
at McDonald's, it's reflected in the tiniest details. Walk into IBM's
New York financial branch. The first thing that greets you is a
massive floor-to-ceiling bulletin board with glossy photographs of
every person in the branch hung under the banner: NEW YORK FI-
NANCIAL . . . THE DIFFERENCE IS PEOPLE.

Watson started an open door policy early and it is still maintained today. Some of his managers used to complain because he so
regularly favored the employees. One former colleague of the senior
Mr. Watson says, in fact, he can hardly ever remember the senior
Watson taking the manager's side. That's the kind of thing that
makes such policies work. They are credible. Managers *do* go to the
trouble of thoroughly checking things out, as in similar open door
situations at Levi, HP, Tandem, and Delta Airlines. It's used.
Things happen.

Thomas Watson, Jr., describes how his father started, foreshad-
owing many continuing IBM policy cornerstones: "T. J. Watson
didn't move in and shake up the organization. Instead, he set out to
buff and polish the people who were already there and to make a

success of what he had. That decision in 1914 led to the IBM policy on job security, which has meant a great deal to our employees." Watson notes that his father even adhered to the policy in the thick of the Great Depression. "IBM produced parts for inventory and stored them. From it has come our policy to build from within. We go to great lengths to develop our people, to retrain them when job requirements change, and to give them another chance if we find them experiencing difficulties in the job they are in." The senior Watson developed his enlightened views under the tutelage of the fabled John Patterson, the founder of NCR. According to Watson, Jr., when others were fighting the union, Patterson was breaking ground by "providing showers on company premises and company time, dining rooms serving hot meals at cost, entertainment, schools, clubs, libraries, and parks. Other businessmen were shocked at Patterson's notions. But he said that they were investments that would pay off, and they did."

Watson followed Patterson's footsteps in many other ways. In his own words, "Almost every kind of fanfare was tried to create enthusiasm. . . . Our early emphasis on human relations was not motivated by altruism but by the simple belief that if we respected our people and helped them to respect themselves, the company would make the most profit."

Detail after detail reinforces the people theme at IBM. A 1940 article in *Fortune* about IBM, then a $35 million company, talks about wholly immaculate factories, the $1-a-year country club for all employees, the IBM song book ("We know and we love you, and we know you have our welfare in your heart"—the "you" in the song being, of course, the senior Watson).

Of the senior Mr. Watson, *Fortune* says he was a "born homilist who began early to confect the altruistic rules of thumb that have since guided his life and policies. He journeys half the time, working 16 hours a day, spending almost every evening at the functions and celebrations at his innumerable employee clubs. . . . He relishes talking to employees, not as a curious supervisor, but as an old friend."

There is not much to add to the early Watson stories, except for the remarkable fact that IBM has stayed about the same. The open door policies, the clubs, the simplicity, the homilies, the hoopla, and the training are as intense in relation to the styles of today as they were fifty or sixty years ago. An IBM executive put it succinctly: "You can foul up on 'most anything and you'll get another chance. But if you screw up, even a little bit, on people management, you're gone. That's it, top performer or not."

Finally, to complete the story on people at IBM, as at other companies, the policies probably would not work if the people way down the organization were not proud of what that organization does. Buck Rodgers, the senior marketing man at IBM, says: "Above all, we seek a reputation for doing the little things well." What IBM stands for, the quality that a Hewlett-Packard or a McDonald's delivers, the ownership of productivity ideas at Dana—in every case the simple pride in what the company does is the keystone for an overarching orientation toward people.

COMMON THEMES

As we step back from the analysis of people and productivity, we find a number of strikingly similar themes running through the excellent companies data. First is language. *The language in people-oriented institutions has a common flavor.* In many respects, form precedes substance. We have seen it happen with some of our clients. Once they start talking the philosophy, they may start living it, even if, initially, the words have no meaning. For example, we doubt that "the HP Way" meant very much to anyone in Hewlett-Packard when the language was first introduced. As time went by, we suspect that the phrase took on deeper and richer meanings in ways that no one would have suspected—not even Hewlett or Packard.

In fact, we doubt that a true people orientation can exist unless there is a special language to go with it. Words and phrases like Family Feeling, open door, Rally, Jubilee, Management By Wandering Around, on stage, and so on—all of these special terms show people in the institutions that the orientation is bone-deep. The Eskimos, unlike the British or the Americans, have many words for various kinds of snow; accurate description of snow conditions is vital to their day-to-day lives, survival, and culture. If an institution is really to be people-oriented, it needs plenty of words to describe the way people ought to treat one another.

Most impressive of all the language characteristics in the excellent companies are the phrases that upgrade the status of the individual employee. Again, we know it sounds corny, but words like Associate (Wal-Mart), Crew Member (McDonald's) and Cast Member (Disney) describe the very special importance of individuals in the excellent companies.

Many of the best companies really do view themselves as an extended family. We found prevalent use of the specific terms "family," "extended family," or "family feeling" at Wal-Mart, Tandem, HP, Disney, Dana, Tupperware, McDonald's, Delta, IBM, TI, Levi Strauss, Blue Bell, Kodak, and P&G. 3M's chairman, Lew Lehr, states the case best:

If you look at the entrepreneurship of American industry it's wonderful. On the other hand, if you look at the paternalism and discipline of the Japanese companies, it's wonderful, too. There are certain companies that have evolved into a blend of those industries, and 3M is one of them. . . . Companies like 3M have become sort of a community center for employees, as opposed to just a place to work. We have employee clubs, intramurual sports, travel clubs, and a choral group. This has happened because the community in which people live has become so mobile it is no longer an outlet for the individual. The schools are no longer a social center for the family. The churches have lost their drawing power as social-family centers. With the breakdown of these traditional structures, certain companies have filled the void. They have become sort of mother institutions, but have maintained their spirit of entrepreneurship at the same time.

And, as Lehr suggests, the family means more than the collection of 3M employees. It includes employees' entire families. One of our colleagues was in the brand-management program at P&G for three months as a summer hire. He recalls that his family still received Thanksgiving turkeys from P&G five years later.

Another of the more striking characteristics of the excellent companies is the *apparent absence of a rigidly followed chain of command*. Of course, the chain of command does exist for big decisions, but it is not used much for day-to-day communication. For information exchange, informality is the norm. People really do wander around, top management is in regular contact with employees at the lowest levels (and with customers), everyone *is* typically on a first-name basis. At the extreme, at wildly successful Activision, a $50 million video games maker growing at 100 percent per year, the phone book is alphabetized by first name!

In trying to explain the phenomenon, a GM manager contrasted one key aspect of the striking difference in performance between two giant plants: "I know this sounds like caricature, but I guess that is how life is. At the poorly performing plant, the plant manager probably ventured out on the floor once a week, always in a suit. His comments were distant and perfunctory. At South Gate, the better plant, the plant manager was on the floor all the time. He wore a baseball cap and a UAW jacket. By the way, whose plant do you think was spotless? Whose looked like a junk yard?"

Wandering around, we suppose, is not for everyone. For many managers, this activity does not come naturally; if they were uncomfortable in such an informal role, their meandering might be viewed as condescending or checking up, and if they used their visits to make on-the-spot decisions, they would be undercutting the chain of command, not simply using the practice as a way of exchanging information. Wandering around and informality, then, probably are not for everybody. On the other hand, without a peripatetic management style, we wonder how vital an institution can really be.

We see important evidences of informality in many other traits. For example, at the excellent companies the physical configuration of facilities is different. Informality is usually delineated by spartan settings, open doors, fewer walls, and fewer offices. It is hard to imagine a free-flowing exchange of information taking place in the palatial, formal, expensively decorated suites that mark so many corporate or even divisional offices.

Hoopla, Celebration, and Verve

Consider this interchange:

General Motors finance staffer: Look, I've been in a foundry, there's no way those guys are going to sing songs like the Japanese or the Tupperware ladies.
Second person (from the Midwest): Caterpillar makes top-drawer equipment. Those people are UAW workers. They don't fool around with hoopla.
Third person (also from the Midwest): I was transferred to Peoria. I didn't work for Cat. But every year they put on a "machine day." All the Cat people and their families go out to the proving grounds and get free beer and sandwiches. Last year's theme was "Cowboys and Indians." All the machines were dressed up in costumes and given names. Then the machines engaged in contests, devouring hills and stuff like that. Everybody lapped it up.
Second GMer: You should see South Gate. The plant manager really enjoys whooping it up. The place became a smorgasbord of signs: "Beat Japan," and the like. Why, they even enticed some Hell's Angels types into singing "God Bless America" at a recent rally.

So Americans don't go in for hoopla? Want more evidence? When Bud Zumwalt was on the navy destroyer where he learned a people orientation, he spent an inordinate amount of time on one element of seeming trivia—changing his ship's voice call sign. He stated the case in a missive to his superiors:

Since recently assuming command of *ISBELL,* this commanding officer has been concerned over the anemic connotation of the present voice radio call. When in company with such stalwarts as *"FIREBALL," "VIPER,"* and others, it is somewhat embarrassing and completely out of keeping with the quality of the sailormen aboard to be identified by the relatively ignominious title *"SAP-WORTH."*

Six months later, after much pulling and tugging, a call-sign change was approved, with drastic subsequent effect. Zumwalt concludes: "The voice call *'Hellcat'* proved immensely popular. Arnold J. Isbell's officers and men proudly wore sleeve patches and baseball cap patches showing a black cat with a forked tail stepping out of the flames of hell and breaking a submarine with its paws. The impact on morale was remarkable."

Kyocera has 2,000 employees in and around San Diego. It's a subsidiary of Kyoto Ceramic, recently named the "foremost company in Japan." Every day at the six U.S. plants, all 2,000 employees assemble first thing in the morning to listen to a management talk about the state of the company. They engage in brisk calisthenics. Management's point of view is "that by doing one thing together each day, it reinforces the unity of the company. It's also fun. It gets the blood up." Top management takes turns making the presentations. Many of the speeches "are very personal and emotional, not approved beforehand or screened by anybody."

At our second meeting with the people at Hewlett-Packard, as we were waiting in the lobby, chief executive Young's voice came over a loudspeaker announcing that quarter's results to everyone in the organization. Young is a soft-spoken individual, but if there is such a thing as quiet cheerleading, that is exactly what Young was doing.

Peter Vaill is a student of "high performing systems"—businesses, orchestras, football teams. Such systems behave, according to Vaill, as self-fulfilling prophecies—something works, for discernible reasons. Then Vaill notes the inevitable emergence of a "private language and set of symbols": people feel "up" because something has worked, and, if allowed, they start to act in a new way. As they act in the new way, more good things happen. "Peak experiences . . . lead members to enthuse, bubble, and communicate joy and exultation. . . . People eat, sleep, and breathe the activity. . . . A Hall of Fame phenomenon arises . . . members acquire an aesthetic motivation." And finally an air of invincibility leads to the same reality.

We haven't the systematic data, so we can't conclude with finality that our excellent companies are far above the norm in the amount of time they spend on training activities. On the other hand, there are enough signs of *training intensity* to suggest that that might be the case. The most visible evidence is the universities—Disney U., Dana U., and Hamburger U., for example. As we saw earlier, IBM invests heavily in training. Caterpillar, similarly, takes its people through extensive training; for instance, all sales engineers spend months at proving grounds learning how the equipment works. Heavy doses of early on-the-job training also mark HP, P&G, and Schlumberger.

An element of Bechtel's on-the-job training may be the most unusual. This company, the builder of $5 billion cities in the Arabian desert, intentionally takes on small, uneconomic projects. "The sole purpose is to provide practical opportunities for fast-track young project managers to cut their teeth on a whole job early," notes a senior executive. (This, by the way, is exactly in the tradition of Alfred Sloan at GM. He almost always put his fast-trackers in the tiny divisions, so they could get an early feel for the full operation and not get lost in the catacombs of a Chevrolet.)

Another striking aspect of the orientation of the excellent companies is the way they *socialize incoming managers*. The first element, of course, is recruiting. The screening is intense. Many of the companies we talked to are known for bringing potential recruits back seven or eight times for interviews. They want to be sure of the people they hire, and they are also saying to would-be recruits, "Get to know our company. Decide for yourself whether or not you can be a good fit with our culture."

Next comes the entry job. This may be the most important element. These companies like to start their aspiring managers in "hands-dirty" positions that are in the mainstream of the business. At HP, according to chief executive Young, "The young MBAs and MSEEs must get immediate experience in new-product introduction. It's a typical starting job. It reinforces the whole concept of bringing new products to market, which is such an important business value to us." Likewise, *Business Week* notes that "Caterpillar has always started its potential managers near the bottom, usually right on the production line. There are no overnight stars in the organization."

The notion of socializing managers by starting them in hands-dirty jobs is strikingly different from what we see in many other large companies. MBAs or other would-be managers, because they are expensive, start in staff jobs and spend years there, never coming to know the reality of the business.

The important result is the realism. Those who start in the company's mainline jobs, the making or selling parts of the business, are unlikely to be subsequently fooled by the abstractions of planning, market research, or management systems as they are promoted. Moreover, their instincts for the business develop. They learn to manage not only by the numbers but also, and perhaps more important, by a real feel for the business. They have been there. Their instincts are good. Bechtel's guiding motto, "A fine feel for the doable," says it well.

The next part in the crucial socialization process is learning through role models—the heroes and the myths. The new recruit learns how to do the job from war stories. At IBM the war stories surround customer service. At 3M the stories are about sometimes failing, but always persisting in pursuit of innovation. At P&G the tales are about quality. HP takes the direct approach by filling its basic indoctrination book, *The HP Way*, with vignettes about those who started at the bottom and made it to the top. HP even systematically collects "HP Way stories" via the suggestion box to add to and revitalize the stock.

Information Availability and Comparison

We are struck by the importance of available information as the basis for peer comparison. Surprisingly, this is the basic control mechanism in the excellent companies. It is not the military model at all. It is not a chain of command wherein nothing happens until the boss tells somebody to do something. General objectives and values are set forward and information is shared so widely that people know quickly whether or not the job is getting done—and who's doing it well or poorly.

Some really do believe in the business of sharing information. A striking example comes from Crompton Corduroy. *Fortune* notes that in one old plant, with the push of a few buttons on a console, machine operators can check on their output and compare it with that of their peers. They do check up on themselves, with no coercion, often cutting a lunch break short to stop at the terminal for a readout. *Fortune* likewise reports on GM's recent decision to disseminate information widely:

Bringing financial information down to the shop floor is a major step in bridging the gap between management and labor; *more than any other single act, it makes the goals explicit and the nature of the partnership concrete* [italics ours]. At the Gear [a huge old Chevrolet plant], managers tell workers the plant's direct labor costs, scrap costs and profit (or loss)—and how these measure up against goals. Not even the foremen would have been privy to such information at GM in the past. The benefits, to GM's way of thinking, outweigh any harm that might come from revealing competitive information.

When Ed Carlson was president of United Airlines, he said: "Nothing is worse for morale than a lack of information down in the ranks. I call it NETMA—Nobody Ever Tells Me Anything— and I have tried hard to minimize that problem." Analyst Richard Pascale observes that Carlson "shared with the field staff confidential daily operating statistics that were previously regarded as too sensitive for the field to handle."

Blue Bell is similarly generous with its comparative productivity information. Individual, team, and unit results are available to everybody. (We have already observed the wealth of information available at companies like Dana.)

Perhaps the prime ingredient in the information-sharing process, a conclusion supported by extensive psychological research, is the nonevaluative nature of the process. There is a fine line here, we will agree. However, what we mean is nonevaluative in a definite sense. Management doesn't browbeat people with numbers. "Superiors" are not telling "subordinates" what to do. On the other hand, the information is evaluative in that it brings to bear a most potent force—namely, peer pressure. For instance, we saw that Dana shoves nothing down the division manager's throat; it just brings that person in for ten days a year, to a pair of five-day Hell Weeks, to swap results on productivity improvement. Intel revealed that its managers swap MBO results—with one another and *weekly*.

A long time ago, the organization theorist Mason Haire said, "What gets measured gets done." He argued that the simple act of putting a measure on something is tantamount to getting it done. It focuses management attention on that area. Information is simply made available and people respond to it. Our favorite story of simple systems, peer pressure, and easy measurement was related to a persistent and pernicious absenteeism problem at one of AT&T's Western Electric plants. Management tried everything; the level of absenteeism wouldn't go down. Finally they put up a huge, visible board with everybody's name on it and posted a gold star next to each name when people came to work. Absenteeism dropped dramatically—almost overnight. Another friend tells of a foreman who started writing production results, after a shift, in chalk on the floor in the machine area. Competition between shifts surfaced and quickly turned intense. Productivity leaped.

All of us, we suspect, are like those Crompton Corduroy machine operators. We sneak by the performance indicator board to find out how we are doing. We respond—more than we likely know or realize—to comparative performance information. The surprise to the unschooled is that we respond better and more strongly if the information is not blatantly evaluative, beating us over the head. Passing the information quietly seems to spur us on to greater effort. Sadly, the excellent companies' policy of making information available stands in vivid contrast to typical management practice, in which so many fear that "they" will abuse the information, and that only competitors will benefit. It's one more big cost of not treating people as adults—or indeed, as winners.

"A man wouldn't sell his life to you, but he will give it to you for a piece of colored ribbon," William Manchester asserts, in describing his World War II experiences as a foot soldier. He echoes a theme that goes back at least to Napoleon, who was a master ribbon-granter. If you want proof of the effect, go back and look through closets and drawers as we recently did. We still have Boy Scout merit badges, trophies gathering dust, and a medal or two from some insignificant ski races held decades ago.

As we did this research, we were struck by the wealth of non-monetary incentives used by the excellent companies. Nothing is more powerful than positive reinforcement. Everybody uses it. But top performers, almost alone, use it extensively. The volume of contrived opportunities for showering pins, buttons, badges, and medals on people is staggering at McDonald's, Tupperware, IBM, or many of the other top performers. They actively seek out and pursue endless excuses to give out rewards.

At Mars, Inc., the extremely successful consumer goods company, every employee, including the president, gets a weekly 10 percent bonus if he comes to work on time each day that week. That's an especially nice example of creating a setting in which virtually everybody wins regularly. As we saw in the early chapters, people like to think of themselves as winners. Even though IBM has a "gold circle" for the top 10 percent of its salesmen, in our minds it is arguably more important that they engage in lots of hoopla surrounding the One Hundred Percent Club, which covers over two thirds of the sales force. When the number of awards is high, it makes the perceived possibility of winning something high as well. And then the average man will stretch to achieve. Many companies do believe in special awards but use them exclusively to honor the top few (who already are so highly motivated they would probably have done their thing anyway). More vital are the ribbons for a good show by the common man. As McPherson states, the real key to success is helping the middle 60 percent a few steps up the ladder.

Our colleague Ken Ohmae described the low state of formal structure in Japan for *Chief Executive:* "Most Japanese corporations lack even an approximation of an organization chart. Managing directors who enjoy great influence on operations seldom appear in the company organization chart.... Many deputies have line responsibilities, but are also absent from these charts. Honda, for instance ... is not clear how it is organized, except that it uses project teams quite frequently." Ohmae also makes the point that in Japan it is unusual to talk about "organization" in any structural sense, or as something different from the total entity itself.

We found *less obvious structuring* and *certainly less layering* at most of the excellent companies. Remember Delta, Dana, and Disney, where interchangeability of people and jobs is a bedrock principle. And Rene McPherson challenges a class at Stanford Business School as he says, "How many layers do you think it takes to run the Catholic Church?" The students think about it and the most they are able to think of is five—the laity, the priest, the bishop, the cardinal, and the Pope. The point is that even in a huge organization like the Church, very few layers are needed to make it work. Excessive layering may be the biggest problem of the slow-moving, rigid bureaucracy. It is done primarily, it sometimes seems, to make place for more managers in an organization. But the excellent companies evidence challenges the need for all those layers. If such layers exist, a kind of Parkinson's law of management structure sets in: extra levels of management mainly create distracting work for others to justify their own existence. Everyone appears busy; but in reality it is simple management featherbedding.

Beyond relatively less structuring and less layering, there is one more vital structural trait that characterizes the excellent companies. We have mentioned it in passing before, but in our minds it is so important in the context of people and productivity that it needs pointed recognition here. The characteristic is: small is productive.

Smallness

A seminal conference on "the creative organization" took place at the University of Chicago over a decade ago. In the midst of the proceedings the following interchange occurred:

Peter Peterson [then president of Bell & Howell]: In industry we are tending to develop a kind of sterile professional manager who has no emotional feelings about the product, who does not "love" the product. He doesn't create anything, but he kind of manages something in a rather artificial way. I heard Ted Bensinger talk about bowling and what he has done for bowling—he has a feeling for this thing, as Ogilvy has a feeling for advertising. I was just wondering whether we have put enough emphasis on our emotional commitment to great cooking, or great advertising, or great something.

David Ogilvy [founder of Ogilvy and Mather]: It's the opposite of detachment.

Gary Steiner [University of Chicago and conference chairman]: The conception that the greatest chef would be the most effective leader in the kitchen is creatively sound, but isn't it restricted to businesses or organizations which would have one clear-cut professional skill? What would you say about General Motors or the University of Chicago, where there is no one, clear-cut professional skill; there is no one dimension?

Ogilvy: It is a bad institution, because it has excessive diversification.

Steiner: How do you make such an institution creative, short of saying: "Let's divide it up?"

Ogilvy: Divide it up.

Peterson: Break up the companies.

The banking industry is undergoing a revolution caused by deregulation. One outcome is the need to offer tailored services such as corporate cash management. These operations have typically been performed in an undifferentiated fashion in the so-called back-office, with unskilled, sweatshop connotations. Barry Sullivan, chairman of First Chicago, in a recent address to the American Banking Association, offered a solution: "What I am really talking about is breaking up the back office factory into separate businesses." Tom Vanderslice, who recently left GE and took over as chairman of GTE, describes a principal objective of his at the new company: "I'm a big advocate of breaking this business—as best we can—into a series of manageable enterprises." A commentator recently said of one of the key ingredients of 3M's continuing success: "As divisions reach a certain size they somehow split, amoeba-like, into smaller, more manageable divisions." And another 3Mer reiterated: "There is only one point. Break it up. Competitive dynamics and efficiencies be damned. It will only stay vital if it's small."

The point of smallness is that it induces manageability and, above all, commitment. A manager really can understand something that is small and in which one central discipline prevails.

More important, even in institutions that employ hundreds of thousands of people, if the divisions are small enough, or if there are other ways of simulating autonomy, the individual still counts and can stand out. We asserted earlier that the need to stick out, to count as an individual, is vital. We simply know no other way individuals can stick out unless the size of units—divisions, plants, and teams—is of human scale. Smallness works. Small *is* beautiful. The economic theorists may disagree, but the excellent companies evidence is crystal clear.

Emerson Electric and Dana are cost-driven companies, and their strategies work. But, at the same time, both hold their division size to well under $100 million. HP and 3M, as we have already noted, strictly limit division size, even though it means overlap and duplication. TI has ninety Product Customer Centers, on average between $40 and $50 million in size.

Johnson & Johnson uses the same magic, even in consumer goods, where large scale is seen as essential by most. With $5 billion in total revenues, J & J, remember, has around 150 divisions—about $30 to $40 million per division on average. Digital employs much the same strategy. "Essentially, we act like a group of smaller companies," says Ted Johnson, vice president for sales and service. At Digital, that means constant reorganization, product-line proliferation and overlap, salesmen out creating "one customer niche after another." People at Digital, and at many of the other excellent companies, regularly lament short production runs, inventory confusion, and sometimes dual coverage of customers. They lament, we'd add, all the way to the bank.

The process of keeping it small can start early. ROLM is a highly successful $200 million telecommunications equipment producer. It does well against giants like Western Electric. The primary reason is that it tailors its problem solving to modest-sized customer segments. In the words of one of its founders, the key to its winning formula is "we continuously divisionalize, and even set up new small buildings for the new units"—and the company grows and grows.

A rule of thumb starts to emerge. We find that the lion's share of the top performers keep their division size between $50 and $100 million, with a maximum of 1,000 or so employees each. Moreover, they grant their divisions extraordinary independence—and give them the functions and resources to exploit it.

For us, the story on plant size was nothing short of astonishing. Repeatedly, we found that the better performers had determined that their small plants, not their big ones, were most efficient. Emerson is the best example. When named as one of the *Dun's Review* "best managed companies," a simple success ingredient was highlighted: "Emerson eschews giant factories favored by such competitors as General Electric. Few Emerson plants employ more than 600 workers, a size at which [Chairman Charles] Knight feels that management can maintain personal contact with individual employees. 'We don't need a 5,000 person plant to get our cost down,' he says, 'and this gives us great flexibility.' Emerson puts heavy stress on those personal contacts with employees."

Blue Bell is number two behind Levi Strauss in the apparel industry. This $1.5 billion giant has managed to stay competitive and profitable, principally on the basis of superb operating skills and low-cost production. In the Blue Bell scheme of things, smallness plays a commanding role. Its chairman, Kimsey Mann, keeps his manufacturing units down to 300 people. This is what he says he gets in return: "A management that is quickly responsive to problems . . . a staff that serves workers." He adds, "We get increased face-to-face contact. Our supervisors have got to know the families, the concerns, of every one of their people." He believes that from the smallness stems creativity and variety. "Who knows the job better than those close to it?" he asks, adding, "In big units, by the time something gets approved, the person who submitted the idea either doesn't remember it or doesn't recognize it as his." In summary, Mann says, "We want a series of plants where a man feels that 'my wife and daughter can work here.' We want every individual to be responsible for the company's image." Mann believes that these traits can exist only if small plant size is maintained.

At Motorola the story is similar. President John Mitchell said simply, "When a plant starts to edge toward fifteen hundred people, somehow, like magic, things start to go wrong." Dana, with its extraordinary record of productivity, tries determinedly to keep its plants down to fewer than 500 employees. Westinghouse is now undertaking a remarkable priority productivity drive; a principal element of the program is a series of thirty to forty small plants. A part of GM's new productivity efforts similarly involves keeping new facility size well under 1,000.

The negative argument appears equally persuasive. A former president of Consolidated Edison said: "In the last decade, the industry [electric utilities] has been euchred into buying individual generating units that are larger than the existing state of construction and operating technology can handle reliably." These words were echoed at one of our briefings by the chief executive of Georgia Power. "Big plants are great," he said, *when they work."* Everyone laughed. He went on to point out that his big plants were shut down much too often, and consequently missed their theoretical potential by a wide margin.

Harvard's Wick Skinner, dean of academic thinkers about production processes, recounts a typical tale, quoted in *Fortune,* revealing what goes on underneath the surface when smallness pays off:

Skinner cites an episode that took place at Honeywell, where he worked for ten years before joining the Harvard faculty. One Honeywell plant was devoted to making gyroscopes for highly specialized scientific and technical use, and fuel gauges for airplanes. The two production lines were intermingled on the factory floor and eventually trouble developed. "Gyroscopes were ten times harder to make," Skinner recalls, "but the Honeywell people were having trouble competitively with the fuel gauges. They did everything to try to figure out why the costs couldn't be kept down. They made accounting analyses, hired an MBA to come on board. Nothing worked, so they decided to get out of the business. Then one of the managers whispered a suggestion to the plant boss, who asked top management for $20,000. . . . They bought plywood and some two-by-fours and walled off a corner of the factory . . . and segregated the workers. Within six months the problem was licked."

The theoretical case is stated succinctly by the British researcher John Child, who reviewed hundreds of studies of the economics of scale: "The economic benefits of large-scale industry have on the whole been considerably exaggerated, especially during the merger and rationalization fever which gripped Europe in the 1960s. The general conclusion which can be drawn from studies of scale in industrial production is that, while there may be important economic thresholds for the small organization seeking to become medium sized, these are not much in evidence for larger units." He goes on to list some reasons. "There is a high correlation between the size of plants and the intensity of industrial unrest, levels of labor turnover, and other costly manifestations of dissatisfaction."

The conclusion we draw from all of this can be defined as a rough guideline. Regardless of industry, it seems that more than 500 or so people under one roof causes substantial and unanticipated problems. More significant, even for the cost-oriented companies, small is not only more innovative but also more productive.

The most significant evidence of small is beautiful is at an even lower level—that of the team, or section, or quality circle. In most of the companies not on our list, the strategic business unit or some other rather large aggregation of human beings is considered the basic building block of the organization. Among our winners the team is the critical factor, regardless of the issue—service, innovation, or productivity. The explanation from an executive at the Bank of America (he heads a large part of the operations organization) is typical:

It's always the same, it seems. We always try to get it exactly right. We always try to optimize. We look for the perfect giant system. I remember when I was in London. I was finally far enough away from the center of things that I could experiment. A long-term problem [endemic to the industry] is getting the operations, the systems, and the credit [lending] people together. We took a small service area. I thought it was a terrific opportunity to experiment with a minicomputer. We could put together a small team to work on the problem. We did, and the results were fabulous. You just can't count the number of ways in which hurdles were overcome. Once that group of ten or twelve people got working together, they were readily able to see each other's contributions. The operations guy had been a shy, bureaucratic sort. But pretty soon it became obvious to his colleagues in systems and credit that he really knew what the hell he was doing. He became the de facto leader of the bunch—even though his grade was substantially lower than that of several of his associates. In a period of only three to four months, they put together a remarkably effective system. It served a discrete bunch of customers. It made money. The morale of the group was sky-high. We ended up using that technique with a great deal of success throughout the London office. It's just amazing how you really can break things up into smallish bits—and get people motivated—if you'll only try.

We've already noted that a disproportionate share of innovation successes in business seems to come from "skunk works," tiny groups that tend to outperform the much larger labs that often have casts of hundreds. We have, now, several score examples of effective skunk works. At Bloomingdale's, 3M, HP, Digital, the entire institution is designed as a wholesale collection of ten-person skunk works. The team is the mainstay of productivity improvement per se. TI insists that virtually every one of its people be on a People Involvement Program team at least once a year. A PIP (or productivity team) is a way of life, arguably *the* way of life at TI.

What are the characteristics of a typical TI team? It is usually limited to eight to ten members, consisting of shop floor people as well as an outside engineer or two, who are typically called in on a voluntary basis by the team. It takes on a limited set of objectives; the point is to come up with something concrete that can pay off in the foreseeable future. The duration is limited to between three and six months. More important, the objectives are always set by the team. Mark Shepherd, chairman of TI, says: "Teams set their own improvement goals and measure their own progress toward these goals. Time after time, team members set what they feel are challenging but realistic goals for themselves, and once the program gets rolling, they find that they are not only meeting but exceeding their goals. This is something that rarely happens if goals are set *for* the team, rather than *by* the team. When we talk about 'improving people effectiveness,' then, we mean giving people these kinds of opportunities to tap their own creative resources." Finally, opportunity after opportunity is used to celebrate team achievements; reviews at all levels are frequent, including a couple of groups quite regularly telling their stories directly to the board of directors.

At TI, each of the 9,000 teams sets its own objectives. At 3M, each of the new product development teams is manned by volunteers, full-timers, headed by a champion. It's the same story for the Dana "store manager" or United Airlines' "station manager." Small size is *the* prime generator of commitment. The analytic model will have no part of such a soft argument, but the empirical evidence is crystal clear. In the words of E. F. Schumacher, "People can be themselves only in small, comprehensible groups."

Philosophy

The excellent companies have a deeply ingrained philosophy that says, in effect, "respect the individual," "make people winners," "let them stand out," "treat people as adults."

As Anthony Jay observes, that lesson (treating people as adults) may have been in front of our eyes for a long time:

One reason why the Roman Empire grew so large and survived so long—a prodigious feat of management—is that there was no railway, car, airplane, radio, paper, or telephone. Above all, no telephone. And therefore you could not maintain any illusion of direct control over a general or provincial governor, you could not feel at the back of your mind that you could ring him up, or he could ring you, if a situation cropped up which was too much for him, or that you could fly over and sort things out if they started to get into a mess. You appointed him, you watched his chariot and baggage train disappear over the hill in a cloud of dust and that was that. . . . There was, therefore, no question of appointing a man who was not fully trained, or not quite up to the job: you knew that everything depended on his being the best man for the job before he set off. And so you took great care in selecting him; but more than that you made sure that he knew all about Rome and Roman government and the Roman army before he went out.

Living by the Anthony Jay principle is the only way a company like Schlumberger can hope to function. The sole way that company can work is to place its faith in its 2,000 well-trained, perfectly socialized young engineers, who are sent to the ends of the earth for months—like the Roman general—and left on their own with only the Schlumberger philosophy and this extensive training to guide them. Dee Hock at Visa summed up the problem when he said, "Substituting rules for judgment starts a self-defeating cycle, since judgment can only be developed by using it."

9

Hands-On, Value-Driven

Let us suppose that we were asked for one all-purpose bit of advice for management, one truth that we were able to distill from the excellent companies research. We might be tempted to reply, "Figure out your value system. Decide what your company *stands for*. What does your enterprise do that gives everyone the most pride? Put yourself out ten or twenty years in the future: what would you look back on with greatest satisfaction?"

We call the fifth attribute of the excellent companies, "hands-on, value-driven." We are struck by the explicit attention they pay to values, and by the way in which their leaders have created exciting environments through personal attention, persistence, and direct intervention—far down the line.

In *Morale,* John Gardner says: "Most contemporary writers are reluctant or embarrassed to write explicitly about values." Our experience is that most businessmen are loath to write about, talk about, even take seriously value systems. To the extent that they do consider them at all, they regard them only as vague abstractions. As our colleagues Julien Phillips and Allan Kennedy note, "Tough-minded managers and consultants rarely pay much attention to the value system of an organization. Values are not 'hard' like organization structures, policies and procedures, strategies, or budgets." Phillips and Kennedy are right as a general rule but, fortunately, wrong—as they are the first to say—about the excellent companies.

Thomas Watson, Jr., wrote an entire book about values. Considering his experiences at IBM in *A Business and Its Beliefs,* he began:

One may speculate at length as to the cause of the decline and fall
of a corporation. Technology, changing tastes, changing fashions,
all play a part. . . . No one can dispute their importance. But I
question whether they in themselves are decisive. I believe the real
difference between success and failure in a corporation can very
often be traced to the question of how well the organization brings
out the great energies and talents of its people. What does it do to
help these people find common cause with each other? And how
can it sustain this common cause and sense of direction through the
many changes which take place from one generation to another?
Consider any great organization—one that has lasted over the
years—I think you will find that it owes its resiliency not to its
form of organization or administrative skills, but to the power of
what we call *beliefs* and the appeal these beliefs have for its people.
This then is my thesis: I firmly believe that any organization, in
order to survive and achieve success, must have a sound set of beliefs on which it premises all its policies and actions. Next, I believe
that the most important single factor in corporate success is faithful
adherence to those beliefs. And, finally, I believe if an organization
is to meet the challenge of a changing world, it must be prepared to
change everything about itself except those beliefs as it moves
through corporate life. In other words, the basic philosophy, spirit,
and drive of an organization have far more to do with its relative
achievements than do technological or economic resources, organizational structure, innovation, and timing. All these things weigh
heavily in success. But they are, I think, transcended by how
strongly the people in the organization believe in its basic precepts
and how faithfully they carry them out.

Every excellent company we studied is clear on what it stands
for, and takes the process of value shaping seriously. In fact, we
wonder whether it is possible to be an excellent company without
clarity on values and without having the right sorts of values.

Led by our colleague Allan Kennedy, we did an analysis of "superordinate goals" about three years ago. (We called it that because that was the way the McKinsey 7-S framework was labeled
at the time. Since then we have changed the term to "shared val-

ues"; but although the wording has changed, we have always meant the same thing: basic beliefs, overriding values.) The study preceded the excellent companies survey, but the result was consistent with what we subsequently observed. Virtually all of the better-performing companies we looked at in the first study had a well-defined set of guiding beliefs. The less well performing institutions, on the other hand, were marked by one of two characteristics. Many had no set of coherent beliefs. The others had distinctive and widely discussed objectives, but the only ones that they got animated about were the ones that could be quantified—the financial objectives, such as earnings per share and growth measures. Ironically, the companies that seemed the most focused—those with the most quantified statements of mission, with the most precise financial targets—had done *less* well financially than those with broader, less precise, more qualitative statements of corporate purpose. (The companies without values fared less well, too.)

So it appeared that not only the articulation of values but also the content of those values (and probably the way they are said) makes the difference. Our guess is that those companies with overriding financial objectives may do a pretty good job of motivating the top fifteen—even fifty. But those objectives seldom add much zest to life down the line, to the tens of thousands (or more) who make, sell, and service the product.

Surprisingly, but in line with Gardner's observation, only a few brave business writers have taken the plunge and written about values. And none of those who have is more articulate than Philip Selznick, whom we introduced in Chapter 4. In *Leadership and Administration,* he talks about values and sketches the leader's hands-on role:

The formation of an institution is marked by the making of value commitments, that is, choices which fix the assumptions of policy makers as to the nature of the enterprise, its distinctive aims, methods, and roles. These character defining choices are often not made verbally, they might not even be made consciously.... The institutional leader is primarily an expert in the promotion and protection of values.... Leadership fails when it concentrates on sheer survival. Institutional survival, properly understood, is a matter of maintaining values and distinctive identity.

Henry Kissinger has stressed the same theme: "The task of the leader is to get his people from where they are to where they have not been. The public does not fully understand the world into which it is going. Leaders must invoke an alchemy of great vision. Those leaders who do not are ultimately judged failures, even though they may be popular at the moment."

In fact, the theoretical case goes deeper. Values are not usually transmitted, as Selznick implies, through formal written procedures. They are more often diffused by softer means: specifically the stories, myths, legends, and metaphors that we've already seen. On the importance of myth as a way of transmitting the value system, Selznick is once again instructive:

To create an institution you rely on many techniques for infusing day-to-day behavior with long-run meaning and purpose. One of the most important of these techniques is the elaboration of socially integrating myths. These are efforts to state, in the language of uplift and idealism, what is distinctive about the aims and methods of the enterprise. Successful myths are never merely cynical or manipulative. . . . To be effective, the projected myth must not be restricted to holiday speeches or to testimony before legislative committees. It requires some interpreting and the making of many diverse day-to-day decisions. The myth helps to fulfill the need. Not the least important, we can hope that the myth will contribute to the unified sense of mission and thereby to the harmony of the whole. In the end, whatever the source, myths are institution builders. The art of creative leadership is the art of institution building, the reworking of human and technological materials to fashion an organism that embodies new and enduring values.

And so, as it turns out, the excellent companies are unashamed collectors and tellers of stories, of legends and myths in support of their basic beliefs. Frito-Lay tells service stories. J&J tells quality stories. 3M tells innovation stories.

Another of our colleagues, John Stewart, is fond of observing: "If you want to know a good company's shared values, just look at its annual report." Sure enough, the annual reports and other publications of the excellent companies make clear what they're proud of and what they value.

Delta Airlines: "There is a special relationship between Delta and its personnel that is rarely found in any firm, generating a team spirit that is evident in the individual's cooperative attitude toward others, cheerful outlook toward life, and pride in a job well done."

Dana: "The Dana style of management is getting everyone involved and working hard to keep things simple. There are no policy or procedure manuals, stacked up layers of management, piles of control reports, or computers that block information and communication paths. . . . The Dana style isn't complicated or fancy. It thrives on treating people with respect. It involves all Dana people in the life of the company."

Caterpillar: "Availability of parts from dealers and from Caterpillar parts distribution facilities combined was at a record high level in 1981." And, "Caterpillar dealers are consistently mentioned by customers as a prime reason for buying Caterpillar products. Many of these dealerships are in their second and third generations of affiliation with the company."

Digital: "Digital believes that the highest degree of interaction in any of its activities needs to be in the area of customer service and support."

J&J: "Back in 1890, Johnson & Johnson put together the original first-aid kit in response to a plea from railroad workers who needed treatment on the scene as they toiled to lay tracks across America. Ninety years later the name Johnson & Johnson is still synonymous with home wound care."

Looking at the examples above, one can understand why reviewers of the excellent companies material sometimes say: "Well, your generalizations are nice, but every company does it a little bit differently." The industry environment, if nothing else, *dictates* that Dana stress themes that are different from, say, those at J&J. Moreover, virtually every one of these companies has had its set of beliefs grooved by a unique individual. Accordingly, each company is distinct; that is why most were so willing to share information with us. Nobody, they believe, can copy them.

On the other hand, we find among the excellent companies a few common attributes that unify them despite their very different values. First, as our original survey intimated, these values are almost always stated in qualitative, rather than quantitative, terms. When financial objectives are mentioned, they are almost always ambitious but never precise. Furthermore, financial and strategic objectives are never stated alone. They are always discussed in the context of the other things the company expects to do well. The idea that profit is a natural by-product of doing something well, not an end in itself, is also almost universal.

A second attribute of effective value systems is the effort to inspire the people at the very bottom of the organization. Suppose that financial objectives were meaningful to 1,000 people, or f ve times that many. Even that impact doesn't go far in today's lai ze enterprise. IBM has more than 340,000 people and Digital more than 60,000. The target of a business philosophy is best aimed, in Kyoto Ceramic chairman Kazuo Inamori's words, at "getting the best from the man with fifty percent ability."

The best service-driven companies clearly understand this, and that is how they are able to deliver so thoroughly on service. But even the good, cost-driven manufacturing companies seem to understand the same thing. Blue Bell, which is particularly cost-and operations-conscious, won't sacrifice quality, especially on its bellwether Wrangler jeans. Chairman Kimsey Mann says unequivocally, "Nobody around here will try to save a dime by taking an extra belt loop off the Wrangler jean." He reasons that the saving of a dime is a target that is important to a bunch of division managers and factory managers. But quality and the image of quality affect everybody—*must* affect everybody—from the newly hired seamstress in the backwoods of North Carolina to Mann himself.

The story about Blue Bell leads us to a third point about the content of beliefs. As James MacGregor Burns has said, "The cardinal responsibility of leadership is to identify the dominant contradiction at each point in history." Any business is *always* an amal-

gam of important contradictions—cost versus service, operations
versus innovation, formality versus informality, a "control" orientation versus a "people" orientation, and the like. It is noteworthy, we
feel, that the value systems of the excellent companies do come
down rather clearly on one side of these apparent contradictions.
The charge that the effective belief systems are mere "boilerplate,"
therefore, is quite unwarranted.

The specific content of the dominant beliefs of the excellent companies is also narrow in scope, including just a few basic values:

1. A belief in being the "best"
2. A belief in the importance of the details of execution, the nuts
and bolts of doing the job well
3. A belief in the importance of people as individuals
4. A belief in superior quality and service
5. A belief that most members of the organization should be innovators, and its corollary, the willingness to support failure
6. A belief in the importance of informality to enhance communication
7. Explicit belief in and recognition of the importance of economic growth and profits.

James Brian Quinn believes that a company's superordinate goals
"must be general. But they must also clearly delineate 'us' from
'them.'" Nothing does it better than "being the best" at something
as is abundantly shown. David Ogilvy notes, "I want all our people
to believe they are working in the best agency in the world. A sense
of pride works wonders." Emerson's Charles Knight adds, "Set and
demand standards of excellence. Anybody who accepts mediocrity—in school, in job, in life—is a guy who compromises. And when
the leader compromises, the whole damn organization compromises." In discussing his service goal for IBM, Thomas Watson, Jr.,
is crystal clear and ambitious: "We want to give the best customer
service of any company in the world."

While the most viable beliefs are soaring in one way or another,
many merely emphasize the details of execution but in a fervent
way. For instance, "We believe that an organization should pursue

all tasks with the idea that they can be accomplished in a superior
fashion," says IBM's Watson. "IBM expects and demands a superi-
or performance from its people in whatever they do. I suppose a
belief of this kind conjures up a mania for perfection and all the
psychological horrors that go with it. Admittedly, a perfectionist is
seldom a comfortable personality. An environment which calls for
perfection is not likely to be easy. But aiming for it is always a goad
to progress."

Andrall Pearson, president of PepsiCo, articulates a similar belief
in improving execution at all levels: "We have learned from experience that the best new-product ideas and competitive strategies are
wasted if we don't execute them effectively. In fact, in our kinds of
businesses, executing extremely well is often more productive—and
practical—than creating fresh ideas. Superb execution is at the
heart of many of our most remarkable successes, such as Frito-Lay
in snacks and Pepsi-Cola in grocery stores."

One theme in the belief structure that came up with surprising
regularity was, in David Packard's words, "innovative people at all
levels in the organization." The excellent companies recognize that
opportunity finding is a somewhat random and unpredictable process, certainly not one that lends itself to the precision sometimes
implied by central planning. If they want growth through innovation, they are dependent on lots of people, not just a few in central
R&D.

A corollary to treating everyone as innovator is explicit support
for failure. Emerson's Charles Knight, J&J's James Burke, and
3M's Lewis Lehr explicitly talk about the need to make mistakes.
Steven Jobs, originator of the hugely successful Apple computer,
which in 1981 approached $750 million in annual sales, says: "I
still make mistakes, a lot. About two weeks ago I was having breakfast with some of our marketing people and I started talking about
all the things that were wrong in a way that none of them could do
anything to resolve. I had about fifteen people really pissed at me
so I wrote them a letter about a week later. In the last paragraph I
told them that I was just in Washington and people were asking me
'How does Apple do it?' I said, 'Well, we hire really great people

and we create an environment where people can make mistakes and grow.'"

The last common theme, informality to foster communications, is at the heart of the HP Way, to cite only one example, and therefore the company makes specific points of its use of first names, managing by wandering around, and its feeling of being one big family. All three amount to explicit direction by the organization's top leadership that the chain of command should be avoided in order to keep communications flowing and encourage maximum fluidity and flexibility.

It is obvious to managers like Thomas Watson, Sr., that values are paramount. But how are they laid down? Here, too, we found striking correlations. As the excellent companies are driven by coherent value systems, so virtually all of them were marked by the personality of a leader who laid down the value set: Hewlett and Packard at HP, Olsen at Digital, Watson at IBM, Kroc at McDonald's, Disney at Disney Productions, Treybig at Tandem, Walton at Wal-Mart, Woolman at Delta, Strauss at Levi Strauss, Penney at J. C. Penney, Johnson at J&J, Marriott at Marriott, Wang at Wang, McPherson at Dana, and so on.

An effective leader must be the master of two ends of the spectrum: ideas at the highest level of abstraction and actions at the most mundane level of detail. The value-shaping leader is concerned, on the one hand, with soaring, lofty visions that will generate excitement and enthusiasm for tens or hundreds of thousands of people. That's where the pathfinding role is critically important. On the other hand, it seems the only way to instill enthusiasm is through scores of daily events, with the value-shaping manager becoming an implementer par excellence. In this role, the leader is a bug for detail, and directly instills values through deeds rather than words: no opportunity is too small. So it is at once attention to ideas and attention to detail.

Attention to ideas—pathfinding and soaring visions—would seem to suggest rare, imposing men writing on stone tablets. But our colleagues Phillips and Kennedy, who looked at how leaders shape values, imply that this is not the case: "Success in instilling values appears to have had little to do with charismatic personality. Rather, it derived from obvious, sincere, sustained personal commitment to the values the leaders sought to implant, coupled with extraordinary persistence in reinforcing those values. None of the men we studied relied on personal magnetism. All *made* themselves into effective leaders."

Persistence is vital. We suspect that is one of the reasons why we see such long periods of time at the helm by the founding fathers: the Watsons, Hewlett and Packard, Olsen, and so on.

Leaders implement their visions and behave persistently simply by being highly visible. Most of the leaders of the excellent companies have come from operational backgrounds. They've been around design, manufacturing, or sale of the product, and therefore are comfortable with the nuts and bolts of the business. Wandering about is easy for them because they are comfortable in the field. These leaders believe, like an evangelist, in constantly preaching the "truth," not from their office but away from it—in the field. They travel more, and they spend more time, especially with juniors, down the line.

This trait, too, is explicitly recognized. Harry Gray of United Technologies writes his own ad copy, says *Business Week*. Gray was trained as a salesman. He says that one of the reasons he does so well (for his Pratt & Whitney Aircraft division) against General Electric's aircraft-engine division is that "I show up in places with the customers where I never see the top management of General Electric." Lanier's chairman, Gene Milner, and its president, Wes Cantrell, are the same. Says Cantrell, "Gene and I were the only president or chairman at last year's major word-processing conference." Or, as his fellow executives have been heard to comment of T. Wilson, Boeing's chief executive, "He's still out in the shop," and, when the occasion arises, "He still makes a few crucial design decisions."

Walking about is an official cornerstone of some policies. Hands-on management at HP was defined thus by R&D executive John Doyle:

Once a division or department has developed a plan of its own—a set of working objectives—it's important for managers and supervisors to keep it in operating condition. This is where observation, measurement, feedback, and guidance come in. It's our "management by wandering around." That's how you find out whether you're on track and heading at the right speed and in the right direction. If you don't constantly monitor how people are operating, not only will they tend to wander off track but also they will begin to believe you weren't serious about the plan in the first place. So, management by wandering around is the business of staying in touch with the territory all the time. It has the extra benefit of getting you off your chair and moving around your area. By wandering around I literally mean moving around and talking to people. It's all done on a very informal and spontaneous basis, but it's important in the course of time to cover the whole territory. You start out by being accessible and approachable, but the main thing is to realize you're there to listen. The second is that it is vital to keep people informed about what's going on in the company, especially those things that are important to them. The third reason for doing this is because it is just plain fun.

David Ogilvy makes much the same point: "Do not summon people to your office—it frightens them. Instead go to see them in *their* offices. This makes you visible throughout the agency. A chairman who never wanders about his agency becomes a hermit, out of touch with his staff."

A leading exponent of the art of hands-on management was United Airlines' Ed Carlson. He describes his approach after taking the helm at United with a background only in the hotel business. United was losing $50 million a year at the time. Carlson turned it around, at least for a while:

I travelled about 200,000 miles a year to express my concern for what I call "visible management." I often used to say to Mrs. Carlson when I'd come home for a weekend that I felt as though I were running for public office. I'd get off an airplane, I'd shake hands with any United employees I could find. I wanted these people to

identify me and to feel sufficiently comfortable to make suggestions
or even argue with me if that's what they felt like doing. One of the
problems in American corporations is the reluctance of the chief
executive officer to get out and travel, to listen to criticism. There's
a tendency to become isolated, to surround himself with people who
won't argue with him. He hears only the things he wants to hear
within the company. When that happens you are on the way to
developing what I call corporate cancer. . . . Let's be specific. Robb
Mangold is senior vice president of United Airlines' Eastern division. If he resented my visits to Boston, LaGuardia, or Newark,
then what I practiced by way of visible management won't work.
These people knew I wasn't out for personal glory. I wasn't trying
to undermine them. What I was trying to do was create the feeling
that the chief executive officer of the company was an approachable guy, someone you could talk to. . . . If you maintain good
working relations with the people in line positions you shouldn't
have any trouble. Whenever I picked up some information, I would
call the senior officer of the division and say that I had just gotten
back from visiting Oakland, Reno and Las Vegas, and here is what
I picked up.

We have talked about the leader as hands-on manager, role model, and hero. But one individual apparently is not enough; it is the
team at the top that is crucial. The senior managers must set the
tone. In instilling critical business values, they have no alternative
but to speak with one voice, as Philip Selznick states: "An important principle is the creation of a homogeneous staff. The development of derived policies and detailed applications will be guarded
by shared and general perspectives." Carlson took this point seriously. When he started those 200,000-mile years, he insisted that
his top fifteen people do the same. During the first eighteen months
of the Carlson reign, all fifteen spent 65 percent or more of their
time in the field.

A practical way in which homogeneity at the top is reinforced is
regular meetings. At Delta Airlines and Fluor, all senior management gathers together informally once a day around the coffee
klatch. At Caterpillar, the senior team meets almost daily without
any agenda just to check expectations and swap agreements about

how things are going. Similar informal rituals occur at J&J and McDonald's.

Obviously, too much homogeneity can lead to a "yes-man" syndrome. But remember Dean Acheson's admonition to Richard Neustadt: Presidents need confidence, not warning. Around the critical business values, lots of yea-saying and reinforcement really do seem to be essential.

A final correlation among the excellent companies is the extent to which their leaders unleash excitement. Remember that HP managers are evaluated in terms of their ability to create enthusiasm. At PepsiCo, president Andy Pearson says: "Perhaps the most subtle challenge facing us in the decade of the eighties is to ensure that PepsiCo remains an exciting place to work." In the same vein, Chuck Knight of Emerson says: "You can't accomplish anything unless you have some fun." And David Ogilvy urged his organization: "Try to make working at Ogilvy & Mather *fun*. When people aren't having any fun, they seldom produce good advertising. Kill grimness with laughter. Maintain an atmosphere of informality. Encourage exuberance. Get rid of sad dogs that spread gloom."

Clarifying the value system and breathing life into it are the greatest contributions a leader can make. Moreover, that's what the top people in the excellent companies seem to worry about most. Creating and instilling a value system isn't easy. For one thing, only a few of all possible value systems are really right for a given company. For another, instilling the system is backbreaking work. It requires persistence and excessive travel and long hours, but without the hands-on part, not much happens, it seems.

10

Stick to the Knitting

Back in the sixties when conglomerates were the rage, Jimmy Ling was down in Washington appearing before an anti-trust committee describing why conglomerates were not in restraint of trade. He put up a chart that said, "How many people in LTV [then Ling-Temco-Vought] know the steel business?" He had just bought Jones and Laughlin. The answer? A big red zero was the next chart in his presentation. I bet today Jimmy Ling wishes the answer to that hadn't been zero, because when Jones and Laughlin went down, Ling lost control of LTV.
— Lew Young, Editor-In-Chief, *Business Week*

Texas Instruments now has a billion dollars in consumer electronics sales, but it hasn't been able, after a decade, to turn a profit on them. TI, moreover, dropped out of the consumer watch business. A major competitor was Casio. Notes one industry observer: "It's really quite simple. No University of Texas–trained electrical engineer is going to come up with the idea that a $18.95 electronic alarm calculator should play Schubert to wake you up in the morning. It's just not constitutionally in the cards."

A *Forbes* article describes Heublein's initial failure in controlling its Colonel Sanders acquisition. Says a Heublein executive: "In the wine and liquor business, it doesn't matter what the liquor store looks like. Smirnoff Vodka doesn't get the blame if the floor is dirty. And you can control your product at the factory. We simply bought a chain of five thousand little factories all over the world, and we didn't have the experience in handling that kind of operation."

There is a huge story waiting to be written, but we won't do more than scratch the surface here. It is a simple fact that most acquisitions go awry. Not only are the synergies to which so many executives pay lip service seldom realized; more often than not the result is catastrophic. Frequently the executives of the acquired companies leave. In their stead remains only a shell and some devalued capital equipment. More important, acquisitions, even little ones, suck up an inordinate amount of top management's time, time taken away from the main-line businesses. Even though Conoco is a relatively close kin, we fully expect that senior Du Pont officers will spend a large part of the next several years trying to learn the oil business as they attempt to master their new acquisition. And that despite—again a typical protestation—the fact that Conoco and Du Pont are "to be run separately."

In the first place, both the qualitative guiding value (most frequently a blend of quality/service, people orientation, and innovation) and the hands-on approach are at war with diversification strategies. The typical diversification strategy dilutes the guiding qualitative theme—in part because the acquired institution undoubtedly has different shared values, but also because themes, even general themes, such as quality tend to lose meaning when the organization strays far afield. Management loses its "feel." It just isn't credible for an electronics executive to talk about quality in a consumer goods business. Hands-on systems of leadership and instilling values thrive only to the extent that they are *totally* credible to those down the line. Credibility is built up almost entirely "because I was there." Without emotional commitment, without understanding of the product, there will be no suspension of disbelief.

Our principal finding is clear and simple. Organizations that do branch out (whether by acquisition or internal diversification) but stick very close to their knitting outperform the others. The most successful of all are those diversified around a single skill—the coating and bonding technology at 3M, for example.

The second group, in descending order, comprises those companies that branch out into related fields—the leap from electric power generation turbines to jet engines (another turbine) from GE, for example.

Least successful, as a general rule, are those companies that diversify into a wide variety of fields. Acquisitions especially, among this group, tend to wither on the vine.

Thus it would appear that some diversification is a basis for stability through adaptation, but that willy-nilly diversification doesn't pay—by any measure.

This is the evidence from contrasting the excellent companies in our survey with those not on the list. Moreover—and surprisingly in light of the amount of merging we observe—virtually *every* academic study has concluded that unchanneled diversification is a losing proposition. For instance, the first systematic study of diversification in American business was published by Michael Gort of the National Bureau of Economic Research in 1962. Gort's data showed a mild positive correlation between the number of products companies added to their offerings from 1939 to 1954 and their growth in dollar sales during the same period. But diversification was *not* positively related to profitability in any way.

The most comprehensive study of diversified companies was conducted by UCLA's Richard Rumelt for his Harvard Business School doctoral dissertation, published in 1974 as *Strategy, Structure, and Economic Performance.* Using a broad sample of large American firms, Rumelt found that those businesses with "dominant-constrained" and "related-constrained" diversification strategies* (two out of eight categories) were "unquestionably the best overall performers." Both strategies are based upon the concept of controlled diversity. In Rumelt's words: "These companies have strategies of entering only those businesses that build on, draw strength from, and enlarge some central strength or competence. While such firms frequently develop new products and enter new businesses, they are loath to invest in areas that are unfamiliar to

---

* The "dominant-constrained" and "related-constrained" categories have diversified "by building on some particular strength, skill, or resource associated with the original dominant activity." The difference between the two is that the dominant-constrained category is very closely related to one skill (e.g., coating and bonding at 3M), while the related-constrained business will include close relationships between businesses, but perhaps different technologies (e.g., a trucking firm that enters the rail business). Ground transportation remains the invariant theme, but there are nonetheless substantial technological differences in the two areas.

management." He adds that these better-performing firms "built their diversification strategies on some central skill or strength." Rumelt's analysis was based on the performance of a valid sample of *Fortune* 500 firms over a twenty-year period.

Rumelt subjected his sample to ten financial analyses, including "annual rate of growth in net sales," "price/earnings ratio of the stock," and "after-tax return on invested capital."

To take a couple of examples, during the fifties and sixties the two top-performing categories averaged 14.6 percent return on equity, 12.4 percent return on capital, and had a price/earnings ratio of 17.5. The worst two categories, including "unrelated passive," had a 10.2 percent return on equity (31 percent less), an 8.6 percent return of capital (30 percent less), and a 14.7 price/earnings ratio (16 percent less). All the findings were statistically significant. Our own extension of Rumelt's findings, conducted by David Anderson, shows that this gap in fact increased markedly in the seventies.

Rumelt's principal finding is clear. Organizations that branch out somewhat, yet still stick very close to their central skill, outperform all others. His analyses do not suggest that "simple is better." An overly simple business—one that depends on a single, vertically integrated combine—is, in fact, invariably a poor performer. We see, rather, that businesses that pursue *some* diversification—a basis for stability through adaptation—yet stick close to their knitting, tend to be the superior performers. Rumelt's model is able to accommodate both the need to adapt (related-businesses outperform vertically integrated single businesses) and the value of managing adaptation around the core skill.

Later studies have confirmed and strengthened both Gort's and Rumelt's findings. In one study published in the *Journal of Finance* in 1975, Robert Haugen and Terence Langetieg tested the popular notion that mergers create operating or strategic synergies which are not present when the firms are owned separately. Their criterion for judging whether mergers produced synergies was return to common stockholders. After evaluating the effects on stock price of fifty-nine nonconglomerate mergers that took place between 1951 and 1968, Haugen and Langetieg concluded: "We detect little evidence of synergism in our sample. . . . Any stockholder could have obtained the same results on his own by combining the shares of the two [merged] companies in the appropriate proportions in his portfolio."

As it turned out, the only clear effect Haugen and Langetieg were able to confirm was an increase in the variation of stockholder returns in the merged firms. In other words, investing in the two companies that chose to merge their assets under a single capital ownership structure was a *riskier* proposition than investing in a pair of companies that elected to remain in their base businesses. This finding, which other researchers have confirmed, casts doubt on one of the primary arguments for mergers—diversifying business risk.

One final study was reported in the *Financial Times* of London in late 1981. Its title suggests a congenial conclusion: "Pioneers—the Anti-merger Specialists." The article, written by the leading economist Christopher Lorenz, concludes that "Pioneering European companies placed more emphasis on specialization than diversification and prefer internal expansion to mergers or takeovers." The study included numerous successful organizations, such as Airbus Industries, Club Méditerranée, Daimler-Benz, and Nixdorf.

We are almost apologetic for subjecting the reader to this onslaught of often arcane analysis. But with merger mania as prevalent as it is, it seems worthwhile to illustrate rather exhaustively the almost total absence of *any* rigorous support for very diversified business combinations.

Case after case demonstrates the difficulty of absorbing the unusual. ITT is a classic example. It was the darling of the stock market for many years; its growth record was enviable. Harold Geneen was able, through sheer intellect and hard work, to keep track of this vast empire. But in many respects it had begun to crumble before he left. The company that Geneen inherited from Colonel Sosthenes Behn, the ITT founder, was largely an international telephone company. As such, its mentality, which subtly persisted under Geneen, didn't fit many of the new acquisitions. One commentator notes: "The tools that it takes to run a phone company in Chile don't add much to the management of Continental Baking or Sheraton Hotels." Eventually even the phone company was threatened, when the market shifted from one of pushing American and European technology to Third World countries (the original ITT magic) to the exotica of electronic switching and satellite communications. In other words, an innovative binge in the telecommunications industry started in the early 1970s, and ITT systems, even in the phone companies, weren't ready for it.

We could go on, but the ITT sorts of difficulties are almost caricatures of problems found, especially, in the unrelated companies. For example, the Transamerica conglomerate, a fair performer, accumulated large losses from their United Artists movie operations. The company, whose basis is in the management of financial institutions (e.g., insurance companies), seemed unable to assimilate the volatility of management in the motion picture industry.

The problem certainly isn't restricted to the avowed conglomerators. We've watched the oil companies in recent years trip over every sort of diversification. Mobil attempted the first big company diversification with the acquisition of Marcor (old Montgomery Ward, plus odds and ends). The oil men didn't understand the retailing business. The result was disaster. Exxon, according to many knowledgeable commentators in the late 1970s, had the elusive big company new venture business licked. Exxon Enterprises was held up as a model to all. *Business Week* even ran a cover story on Exxon's presumed role as a future giant competitor to AT&T and IBM in the communications business. But Exxon Enterprises has come on hard times, to put it mildly.

Exxon's experiment worked well when it was tiny. The entrepreneurs and their small enterprises that Exxon acquired were by and large allowed to do their own thing. They had some singular successes—so much so that they became, unfortunately, quite visible to Exxon corporate management. So Exxon, now shifting to the traditional failure-inducing route of new venture activity in large corporations, decided to "help out." It quickly rationalized the businesses, sticking the entrepreneurial parts together into "logical" groupings to gain "market synergy," naturally. It also provided financial "help." A senior treasury official from corporate headquarters moved in to help out the little enterprises with bookkeeping. The rationalization was very premature for an entrepreneurial enterprise. The original entrepreneurs deserted the ship. What was left was a slow-moving infrastructure in a fast-moving market.

Even lesser forays afield, however, show the difficulty of absorbing the unusual. For example, General Electric had tremendous success in entering the aircraft-engine business, and Westinghouse had a major failure. Westinghouse's failure came from a belief that "a turbine is a turbine." They tried to manage the aircraft-engine business within the power-generation organization. It turns out that tolerances and the like in aircraft turbines are radically different from those in turbines used in electrical-power generation. Gerhard Neumann at GE and Jack Parker had recognized this. They kept GE's fledgling aircraft-engine business outside the power-generation old boy network. They put it in a different physical location, Lynn, Massachusetts. They went out and hired specialist engineers who understood the constraints of aircraft-turbine design and production. They succeeded beyond GE's wildest dreams; Westinghouse failed.

An analogue to the GE/Westinghouse tale is readily observable today in the electromechanical to electronic transformation. The thought process involved in electromechanical businesses bears only a modest resemblance, apparently, to that of electronics. So we find that none of the leading vacuum-tube producers in 1965 (the top ten) was among the leading semiconductor producers in 1975, only ten years later. The giants who have fallen by the wayside, unable to make the intellectual leap, included management paragons of an early era, such as GE, RCA, and Sylvania. Two of those three electromechanical stars, GE and RCA, had similar disasters when they tried to get into computers. In theory, the gap should have been just a few steps. After all, an electron is an electron. *But, in practice, a few steps almost always represent a giant leap for a big business.*

If the aircraft-turbine versus power-generation-turbine tale at GE/Westinghouse sounds like a story of closely linked intellectual skills, how about the National/Pan Am merger? Hardly a leap of faith! The exact same industry. Except that it didn't turn out that way. Pan Am, the giant in the international air-passenger service business, seems to have misunderstood National's domestic route structure and capabilities as a would-be feeder airline for Pan Am. Pan Am, it appears, bought a bunch of National DC10s, which turned out to be horribly mis-sized for the combined route structure. Pan Am could have inherited a problem that threatened the life of one of America's historically important companies.

The crucial question, then, is: How have the excellent companies avoided these traps? The answer is simple. The excellent companies don't test new waters with both feet. Better yet, when they stuck a toe in new waters and failed, they terminated the experiment quickly. As a general rule, the top performers moved out mainly through internally generated diversification, one manageable step at a time.

We found that the excellent companies act as if they believe philosophically in what the academics are saying about diversification. As we mentioned, Robert Wood Johnson, J&J founder, said as his prime parting advice to his successor-elect, "Never acquire any businesses that you don't know how to run." Or as Ed Harness, former P&G head, said, "This company has never left its base. We seek to be anything but a conglomerate."

Yet the excellent companies are far from simple. 3M has well over 50,000 products, and introduces well over 100 substantially new ones every year. Only the company's basic coating and bonding technology acts as a common thread. The attributes that hold 3M together surpass those of other companies in many ways, but are at the same time very typical. The top leadership is made up principally of chemical engineers, almost all of whom have spent time in the sales function, working on practical applications. The company's essential skill—customer problem solving for industrial niches based on 3M's technology—is thus enshrined in the top management structure.

The focused disciplinary litany à la chemical engineering at 3M is found in many of the excellent companies. You almost have to be an electrical engineer to make it at HP; or a mechanical engineer to succeed at Fluor or Bechtel; an aeronautical engineer to move up at Boeing; an ex-product manager at Procter & Gamble; or an ex-salesman at IBM. These are the only candidates for the top. So either the specific technological discipline or extensive schooling in the major functional business discipline is highly "over-represented" in senior management in such leading organizations.

The stories mount up:

- *Boeing:* "Observers," notes *The Wall Street Journal,* "say Boeing's strength comes from its almost singular devotion to the commercial airline market where it derives almost 90% of its revenue. 'The other guys are too busy chasing military bucks,' says one airline official. 'At Boeing the airlines come first.'"
- *Fluor:* Chairman Bob Fluor comments, "We can't be everything to all people."
- *Wal-Mart:* Wal-Mart's extraordinary record of growth has come from an overpowering niche strategy. It has stayed in a dozen states. Sticking to what it knows best, it overpowers better-financed and more deeply experienced organizations such as K mart in its chosen area.
- *Deere:* Deere's president, Robert Hanson, states: "We're sticking to the customers we know." *Forbes* adds: "For years Deere has outperformed its archrival, International Harvester. Harvester's allegiance was divided between its truck business and farm machinery. Deere, by contrast, knew what its business was, who its customers were, and what they wanted."
- *Amoco: The Wall Street Journal* contrasts the successful strategy of Amoco with its competitors: "'The wisdom guiding this year's mammoth oil acquisitions is that it's cheaper to buy someone else's reserves than to develop them internally. But at Standard Oil Co. (Indiana), we don't believe that, not for ourselves,' says chairman John Swearingen."

Virtually all the growth in the excellent companies has been internally generated and home-grown. The few acquisitions followed a simple rule. They have been small businesses that could be readily assimilated without changing the character of the acquiring organization. And small enough so that if there is failure, the company can divest or write it off without substantial financial damage.

A few companies *have* thrived on growth via acquisition, but via a "small is beautiful" strategy—notably Emerson and Beatrice Foods. Respectively $4 and $10 billion giants, they have grown mainly by adding $20 to $50 million businesses. They don't believe, apparently, in the oft-cited wisdom that "A $500 million acquisition is no tougher to assimilate than a $50 million one, so make one deal instead of ten." Emerson and Beatrice scan constantly. And add by small bites. In those instances in which the small acquisitions have new strengths (e.g., disciplinary skills) to add to the core businesses, they let it happen naturally through informal interchange and natural diffusion. The strengths are allowed to seep into the corporation.

Similarly, one finds constant *tiny* acquisition activity at an HP or a 3M. Acquisitions are usually $1 to $10 million businesses; they are often transparent efforts to acquire a window on a new skill, but at a size that is manageable enough to allow early, painless integration. It can amount to buying up a few employment contracts. So, small acquisitions can work, or even major new strategic thrusts based on numerous small acquisitions.

This, in brief, then, is the excellent companies' story. They do acquire; but they acquire and diversify in an experimental fashion. They buy a small company or start a new business. They do it in manageable steps . . . and clearly contain the risks. And are willing to get out if it doesn't work.

Thus, we would expect to find—and do find—numerous tales of modest-sized failures among the excellent companies. Even some immodest-sized ones! They demonstrate that even among the best performers, forays of more than moderate reach quite frequently lead to problems.

In fact, the excellent companies may even have special difficulties in reaching far afield. For the cultures that lead them to superior performance do so by emphasizing reasonably narrow business competences. There is no better definer and penetrator of modest-sized industrial niches (up to $100 million or so) than 3M. Yet there are apparently some things that even a 3M cannot do.

Here are a few examples of ill-advised forays among our top performers.

3M: 3M has not been able regularly to exploit its technical muscle in consumer goods businesses. Analysts have suggested that the atomization of 3M (and the personalized industrial selling discipline) inhibits the vast promotional efforts and bets on a small number of products that mark consumer goods marketing efforts. So while 3M has had some successes, by and large its consumer goods businesses are not as profitable as the rest of its portfolio. At the other extreme, 3M has had some difficulties of late in moving into the "office of the future" arena. The problem is much the same as in consumer goods. The more sophisticated office-of-the-future products are "systems products." Once again, the extraordinary autonomy of the 3M divisions is inconsistent with the close divisional linkages often required for complex systems product development and selling efforts.

Hewlett-Packard: We noted earlier that HP had early difficulties in marketing its hand-held calculators. HP's story is closely akin to 3M's. In the instrument and electronics business, HP knows how to serve the professional customer, the modest-sized niche. The $9.95 calculator consumer is beyond HP's ken. Similarly, HP had a disaster with an electronic wristwatch. The mistake was understandable. HP believed that the electronic part was so exotic that the average consumer would look at it as something very special. They were wrong, and the TI $8.95 wristwatches knocked them out of the water. (Company after company in the industry has had difficulty making the transition to the mass electronics arena. National Semiconductor, a mass marketer of chips, also fell flat on its face when it got near the consumer—again with watches. So did Fairchild Semiconductor.)

Texas Instruments: We noted earlier that TI has difficulty dreaming up alarm clock calculators that play Schubert—something that doesn't handicap the consumer-conscious Japanese electronic engineer. Accordingly, TI has generally had trouble with its electronics business in the consumer areas. Parts of the remaining consumer electronics business are profitable, machines such as

Speak 'N Spell. But one suspects that these are profitable because the technology is still "exotic enough" to afford a comparative edge. When the solid-state speech circuits become common currency, as are the chips that run the watches and hand-held calculators, TI may once again fold in the face of the Japanese challenge.

Procter & Gamble: One commentator notes that P&G has trouble with the purely trendy, fashion-conscious twists and turns of the consumer goods business. P&G has always stood, above all, for quality. It doesn't introduce a new product or a reformulation of an old one unless it knows it has a distinctive edge. Thus, one long-time observer points out that P&G had difficulty, in the toothpaste business, in adding the little green stripes that have sold so well in recent years. Additionally, it has expended hundreds of millions of dollars trying to launch Pringles potato chips, despite the effort's continuing failure. Again, the story reflects the company's firm attachment to quality as opposed to gimmicks. The Pringles chip is a classic P&G idea, notes one competitor: a uniform chip put into a nice tidy can. It meets the P&G sense of quality, though from a consumer-taste standpoint it is an apparent bomb.

Sears: For years and years, Sears, Roebuck thrived under the flag of "quality at a decent price." It perceived a need to go up-scale—and failed badly. The business writer Gordon Weil concludes: "Imagine McDonald's introducing a sirloin steak, raising the price of its Big Mac, and withdrawing its plain hamburger. That was Sears' growth strategy. In short, Sears was trying to do two things at once."

All of these examples cast suspicion, we think, on the viability of wide-ranging conglomerations of the kind so enthusiastically pursued in the 1960s. And now we think we detect the beginning of a possible counterrevolution. For example, *The Wall Street Journal* in late 1981 headlined an article: "Colgate Works Hard to Become the Firm It Was a Decade Ago: It Shucks Many Acquisitions and Moves to Strengthen Its Traditional Products":

Keith Crane's [CEO] predecessor, David Foster, had tried, unsuccessfully, to propel Colgate out from under the shadow of the industry's giant, Procter & Gamble, through additions such as sports, food, and apparel companies. In that shopping spree, Colgate man-

aged to buy a lot of trouble. It was turned into a company sapping
the profits of its traditional lines to acquire other businesses already
past their peak profitability. . . . Mr. Crane imposed a searing re-
trenchment. He severed most of Mr. Foster's $935 million of acqui-
sitions—at a cost of at least $96.5 million in reported write-offs. . . .
He also reorganized management, revised advertising budgets and
moved to strengthen basic product lines with a new emphasis on
production and profitability. . . .

Mr. Foster had said, "One of the most exciting and productive
aspects of our company's new direction is the increasing emphasis
on developing new product categories distinct from our traditional
product line, in which market growth is generally limited to the
growth of the population."

"They were all vanity acquisitions," says an advertising executive. "After Kendall [a medical supplies company] and Riviana [a
rice producer], the rest was junkola." Moreover, Colgate's efforts
to introduce new products generally foundered during the Foster
regime. Frequently, the company took a short cut in rolling out new
merchandise; it acted as a mere distributor rather than developing
its own products. . . . "These are mistakes that P&G would never
make," a former consultant to Colgate says. "You learn in this
market to make simple and functional products. Colgate went for
additives."

This story, excepting perhaps the possible happy ending of a roll-
back, is increasingly familiar. The company decides it is in a slug-
gish business. It determines to move afield. It doesn't know what it
is buying. It buys companies at or past their peak. Moreover, it
doesn't understand them (e.g., vanity acquisitions). Finally, and
most devastating of all, the effort and attention going into the management of the new acquisitions sap the vitality of the already
shaky core business. New products (line extensions or reformula-
tions of the old products) are given short shrift or subjected to
"short cuts," as in the Colgate case. And the downhill spiral is
under way.

But the successful de-acquisitioning ending is becoming at least a
little more commonplace. On a single day in late 1980, the *New
York Times* carried stories of divestitures by Litton, Textron, and

GAF. One comes across such news quite frequently today. For example, a 1981 *Business Week* article noted that ITT had divested itself of thirty-three businesses since 1979; a 1981 *Fortune* story commented that Consolidated Foods had sold off fifty businesses in the last five years; a *New York Times* article discussed Great Britain's GEC, now on a de-merger binge (the chairman is quoted as remarking, "One can say that turbines are linked to switchgear, which is linked to transformers, which are linked to control gear, which is linked to lamps. But lamps have no direct links with turbines"); *Forbes* noted, again in 1981, that since 1972, CEO John Hanley of Monsanto had written off more than $800 million in sales "to get back to basics"; and the same magazine observed that Litton's sell-off was being achieved so that the company can "return to the common thread of technology."

This evidence suggests perhaps less than a groundswell, especially in an environment in which Reagan's FTC is giving clear signals that mergers of almost any sort are okay. But *any* "back to basics" move is, according to the studies we have reviewed and the excellent companies' message, good news indeed.

11

Simple Form, Lean Staff

Along with bigness comes complexity, unfortunately. And most big companies respond to complexity in kind, by designing complex systems and structures. They then hire more staff to keep track of all this complexity, and that's where the mistake begins. The solution just doesn't go well with the nature of people in an organization, in which things need to be kept reasonably simple if the unit is truly to pull together. The paradox is clear. On the one hand, size generates legitimate complexity, and a complex systems or structural response is perfectly reasonable. On the other hand, making an organization work has everything to do with keeping things understandable for the tens or hundreds of thousands who must make things happen. And that means keeping things simple.

Our favorite candidate for the wrong kind of complex response, of course, is the matrix organization structure. The matrix is a perfectly plausible idea. As soon as an organization engages in manifold businesses, and is forced to move from that simplest of all forms, the functional structure—finance, sales manufacturing—the dimensions around which it might decentralize become many. The company could organize around product groupings. It could organize around market segments. It could organize around geographic areas where it has plants or sales offices. And, to be sure, the basic functions of finance, sales, and manufacturing don't go away. But if one tries to take all these into account in a formal organization structure, one has at least a four-dimensional matrix, which is a logistical mess.

The dilemma is that the world really is at least that complex. Thus matrix conditions are present in every large organizational setting. The dilemma gets even more complicated when you start adding other sensible ways of organizing—for example, the temporary devices like project centers. What is a manager to do?

Some companies have decided that, even though they couldn't take all possible dimensions for the matrix formally into account, they could at least use some—and come up with a formal structure that gives equal authority over departments or divisions to both product managers and functional managers, such as engineering, marketing, and production. But even that is very confusing, we find. People aren't sure to whom they should report for what. The most critical problem, it seems, is that in the name of "balance," everything is somehow hooked to everything else. The organization gets paralyzed because the structure not only does not make priori‑ ties clear, *it automatically dilutes priorities.* In effect, it says to people down the line: "Everything is important; pay equal attention to everything." The message is paralyzing.

Virtually none of the excellent companies spoke of itself as having formal matrix structures, except for the project management companies like Boeing. But in a company like Boeing, where many of the matrix ideas originated, something very different is meant by matrix management. People operate in a binary way: they are *either* a part of a project team and responsible to that team for getting some task accomplished (almost all the time), *or* they are part of a technical discipline, in which they spend some time making sure their technical department is keeping up with the state of the art. When they are on a project, there is no day-in, day-out confusion about whether they are really responsible to the project or not. They are.

Just to be clear, we are not overly concerned about the organizational form that a few of the early users of the technique—such as Boeing and NASA—called "matrix" management. The key to making these systems work is the same key that makes structures work in the rest of the excellent companies. *One dimension—e.g., product or geography or function—has crystal-clear primacy.* We are concerned about the way that concept has gotten bastardized, so that trying to sort out who is responsible for what, and under what circumstances—and, incidentally, "which boss do I report to on this one or do I keep everyone informed?"—and keeping it all straight become nearly impossible. This breeds staffers who gain and retain substantial power by ensuring that everything stays complex and unclear (i.e., the staff becomes the umpire at the matrix "crossover" points, where, say, product and function clash).

How have the excellent companies avoided this? The answer is, in a number of ways, but underlying it all is a basic simplicity of form. Underpinning most of the excellent companies we find a fairly stable, unchanging form—perhaps the product division—that provides the essential touchstone which everybody understands, and from which the complexities of day-to-day life can be approached. Clarity on values is also an important part of the underlying touchstone of stability and simplicity as well.

Beyond the simplicity around one underlying form, we find the excellent companies quite flexible in responding to fast-changing conditions in the environment and in dealing with the issues posed by the ubiquitous presence of matrix-like conditions. Because of their typically unifying organization theme, they can make better use of small divisions or other small units. They can reorganize more flexibly, frequently, and fluidly. And they can make better use of temporary forms, such as task forces and project centers. They are rearranging the ornaments, but seldom the branches. (Of course, other attributes help keep the organization fluid; e.g., personnel policies that assure security and make people in the company less dependent on the particular organization box they live in.)

The most common simple form we found was the product division. Several companies, however, have avoided the matrix simply by maintaining something close to the old functional form. Companies like Frito-Lay and Kodak are close to that. Finally, others, including McDonald's, are simply organized around their restaurants, stores, boutiques, or factories as the basic building block.

A wonderful example of simplicity in form despite size is Johnson & Johnson. This company represents an extreme of keeping the structure simple, divisionalized, and autonomous. As we've seen, J&J is a $5 billion company broken up into 150 independent divisions, average size just over $30 million. The divisions are each called "companies," and each of them is headed by a "chairman of the board." The companies are aggregated into eight groups of up to twenty companies each, and the companies in each group have either a geographic or a product similarity. Even though none of the companies is truly independent in the sense of having stock of its own, the "boards of directors" are active and buffer the divisions from unwanted (and usually unneeded) corporate interference. A *Wharton Magazine* commentator adds, " [J&J's] central staff is small; no specialists continually travel among the subsidiaries as at General Electric."

For its consumer business, which is the source of about 40 percent of J&J's sales and profits, the organization is straightforward: There are over fifty-five consumer product divisions; each is responsible for its own marketing, distribution, and research. This flies in the face of conventional wisdom, which asserts that consumer market dominance requires large-scale activity. The number could be less and the size of each unit bigger, but they're not for a reason, says chief executive James Burke, in a theme bearing eerie resemblance to that in many of the other excellent divisionalizers:

We have periodically studied the economics of consolidation. Let's just take our consumer business and consolidate the distribution network. There would be some dollar efficiencies on paper. But we say to ourselves that these efficiencies would have to be enormous before we'll go with them, because we believe that if the manager of a business can control all aspects of his business it will be run a lot better. And we believe that a lot of the efficiencies you are supposed to get from economies of scale are not real at all. They are elusive. Once you get your big monster going, you're going to create inefficiencies that you don't know are there. And if the management does see them, it won't be aggressive in rooting them out, because it doesn't have control of them.

The simplicity of form forced by a philosophy like that is closely akin to other stirring examples from our research. Among the supports that make these product division–as–building–block structures work are:

1. Extraordinary divisional integrity. All the main functions, including product development, finance and personnel, are in each division.

2. Constant hiving off of new divisions and rewards for so doing. J&J's 150 divisions are up from only 80 ten years ago. (This point is a fascinating one for us because so many companies reward the reverse, empire builders who create large, layered monoliths.)

3. A set of guidelines that describes when a new product or product line automatically becomes an independent division, e.g., at about the $20 million level at 3M.

4. Shifting people and even products or product lines among divisions on a regular basis and without the acrimony that this would create in most companies.

Interestingly, the simple form is not limited to companies that make a specialty of penetrating discrete, modest-sized niches, or of creating such niches for themselves, companies such as J&J, HP, Emerson, Digital, Dana, and 3M, although the simplicity of the small product division may be most obvious there. Regardless of industry or apparent scale needs, virtually all of the companies we talked to placed high value on pushing authority far down the line, and on preserving and maximizing practical autonomy for large numbers of people. Those things cannot occur without a fairly simple underlying form. They certainly cannot be accomplished within a formal matrix structure.

In a funny sort of way, simplicity in basic structural arrangement actually facilitates organizational flexibility. It seems that because the basic form is clear, flexibility around the base structure is made easy. The excellent companies, as we've seen, do make better use of task forces, project centers, and other ad hoc devices to make things happen. The excellent companies also *appear* to be reorganizing all the time. They are; but most of the reorganization takes place around the edges. The fundamental form rarely changes that much. Boeing is an interesting case. Often the project structure, with some justification, is considered the forerunner of, or a principal example of, the formal matrix. But in reality, each project manager at Boeing retains extraordinary autonomy. And Boeing is proud of its ability to pull people from several layers down in the technical structure and put those individuals in charge of major projects, often with higher-salaried, more senior people reporting to them.

It appears to us that there is only one crucial concomitant to the excellent company's simple structural form: lean staff, especially at the corporate level. As we have shown before, these two attributes seem deeply intertwined and self-fulfilling. With the simple organizational form, fewer staff are required to make things tick.

Indeed, it appears that most of our excellent companies have comparatively few people at the corporate level, and that what staff there is tends to be out in the field solving problems rather than in the home office checking on things. The bottom line is fewer administrators, more operators. As a result, we coined our rough "rule of 100": With rare exception, it seems there is seldom need for more than 100 people in the corporate headquarters.

- Emerson Electric has 54,000 employees and makes do with fewer than 100 in corporate headquarters.
- Dana employs 35,000 people and cut its corporate staff from about 500 in 1970 to around 100 today.
- Schlumberger, the $6 billion diversified oil service company, runs its worldwide empire with a corporate staff of ninety.

McDonald's numbers are similarly low, following Ray Kroc's longstanding dictum that we have mentioned: "I believe that 'less is more' in the case of corporate management." At $1 billion Intel, there is virtually *no* staff. All staff assignments are temporary ones given to line officers. At $2 billion Wal-Mart, founder Sam Walton says that he believes in empty headquarters rule: "The key is to get into the stores and listen." And at Heinz's successful $1 billion subsidiary Ore-Ida, one of the most thoughtful strategic plans we've seen is put together by the president, with the only staff help coming from his secretary and part-time work from his department and division manager. He has no staff, let alone planning staff.

The same extraordinary rule holds for some of the top-performing smaller companies. ROLM, for instance, manages a $200 million business with about fifteen people in corporate headquarters. When Charles Ames took over as head of $400 million Acme Cleveland, he was appalled by the amount of staff. In the space of a few months, he had reduced the corporate headquarters from 120 to 50.

The absolute numbers in these instances are impressive. But at least as important are the kinds of people who are on these staffs. First, what functions have to be retained for corporate? The answer, in many of the excellent companies, is practically none. Product development, usually a corporate or group activity, is wholly decentralized to the divisions in J&J, 3M, HP, and others. Dana takes pride in the decentralization of such functions as purchasing, finance, and personnel—all the way down to the factory level. Strategic planners certainly have a corporate function. Yet Fluor runs its $6 billion operations with three corporate planners. 3M, HP, and J&J have had no planners at the corporate level. Virtually every function in the excellent companies is radically decentralized, down to the division level, at least.

Bechtel has an active research function, yet insists even in the specialized area of research that virtually everyone move into a line operation. Many of its research staffers come in from a line operation and then go again. At IBM, management adheres strictly to the rule of three-year staff rotation. Few staff jobs are manned by "career staffers"; they are manned by line officers. Moreover, those who do get into the rotation on the corporate staff know that within three years they are going back out to the line again. It is a marvelous check on the invention of complex systems. If you know you are going to become a user within thirty-six months, you are not likely to invent an overbearing bureaucracy during your brief sojourn on the other side of the fence. Digital and 3M follow virtually the same rules. Digital and 3M staffs, with the exception of a few legal and financial types, are almost always from the line—and will head back out to the line again.

A related correlation is to be found in the sorts of slots—i.e., number of hierarchical levels—that the staffers fill. Decades ago Americans got hooked on the notion of optimal spans of control. We conventionally believe that no one can control more than five to seven people. The Japanese think that is nonsense. At one bank, more than several hundred branch managers report to the same person. The flat organization *is* possible. One of the biggest contrasts between Japanese and American corporations, in fact, is in number of middle management levels. As we've seen, whereas there are five levels between the chairman and the first-line supervisor at Toyota, Ford has over fifteen.

Now take ex-UAL chairman Ed Carlson's hourglass theory. Middle management in most organizations really has little role beyond "make work" activities, such as stopping ideas coming down and stopping ideas going up. Middle managers, says Carlson, are a sponge. Hands-on management becomes a lot more workable when there are fewer people in the middle.

The numbers in many companies—both levels and employees—are staggering. Ford over the last twenty-four months, in an effort to become more competitive with the Japanese, has cut more than 26 percent of its middle management staff; President Donald Petersen believes this is only the beginning. Reductions in the neighborhood of 50 percent, or even 75 percent, in levels and bodies are not uncommon targets when businessmen discuss what they could honestly do without.

A "FORM" FOR THE FUTURE

What precisely is the organizational character that seems to work best? Each of several organizational forms has major strengths and major weaknesses. Let's now consider them again:

- The functional organization, typical of old-line consumer products firms, is efficient and does the basics well; it is not particularly creative or entrepreneurial, does not adapt quickly, and is especially apt to miss big changes.
- The divisional organization, of which Sloan's GM was prototypical, can do the basics adequately and it is usually more adaptive than the functional organization. But the divisions invariably get too big, and big divisions suffer from all the problems of oversized functional structures. Moreover, divisional organizations often drift into a hodgepodge of centralized and decentralized activity.
- The matrix response to manifold pressures on various fronts—to the overcomplexity of divisional structures, in fact—is in tune with today's realities. On the other hand, it virtually always ceases to be innovative, often after just a short while. It has particular difficulty in executing the basics (the authority structure is uniquely weak). It also regularly degenerates into anarchy and rapidly becomes bureaucratic and noncreative. The long-term direction of the matrixed organization is usually not clear.
- The adhocracy responds to multiple pressures without inducing new permanent bureaucracy. But it, too, can become anarchic if all parties are chasing temporary problems and the basics are ignored (e.g., old functional strengths are sapped as all play Musical Chairs on temporary project teams).
- The missionary "form," as Henry Mintzberg calls it, like that at McDonald's, provides stability via nonstructural means. If it is matched, as in theory it should be, with plenty of experimentation within the value set (and if the value set is appropriate), all can go well. But, as is true of all dogma-based "structures," it can become quite narrow-minded and rigid—even more so than the functional form.

Considering our findings, we would like now to propose a hybrid alternative to all of these forms, to describe the properties of a potential "structure of the eighties," one that will respond to the three prime needs revealed above: a need for efficiency around the basics; a need for regular innovation; and a need to avoid calcification by ensuring at least modest responsiveness to major threats. Accordingly, we think of the resultant structural "form" as based on "three pillars," each one of which responds to one of these three basic needs. To respond to the need for efficiency around the basics there is a stability pillar. To respond to the need for regular innovation, there is an entrepreneurial pillar. And to respond to the need for avoiding calcification, there is a "habit-breaking" pillar.

In the diagram on the following page, the *stability pillar* is based on maintaining a simple, consistent, underlying form, and on developing and maintaining broad yet flexible enduring values. We believe the simple underlying form should generally be the product-based division, that the old, simple, divisionalized organizational structure is probably the best form around—now and for the future. This betrays our clear bias toward the product side, and against the matrix. Everything we have been talking about—entrepreneurship around product and service, a love for the product, quality, a focus on operations and productivity through people—leads us typically to a product or market bias. It is simple, clearer, more direct, more tangible, more honest.

The second feature of the stability pillar is the underlying value system, which encompasses the missionary "form." It may seem strange to talk about values under the heading of organizational structure, but, remember, structure, most broadly defined, *is* communications patterns. When we think about stable forms at IBM, HP, or Dana, for example, we appreciate instantly the need for and desirability of a stable value system.

The heart of *the entrepreneurial pillar* is "small is beautiful." And the way to stay small is constantly to hive off new or expanded activities into new divisions. In this scheme of things, smallness is viewed as a requisite for continual adaptiveness. The cost is occasionally some efficiency; but as we have seen time and again, the efficiency advantage is usually vastly overrated.

Other features of the entrepreneurial pillar are measuring systems and the use of corporate staff elements. When the form is simple and does not depend on vast integrating systems, one can survive with simpler systems and smaller staffs to run the organiza-

THE THREE PILLARS OF THE "STRUCTURE OF THE EIGHTIES"

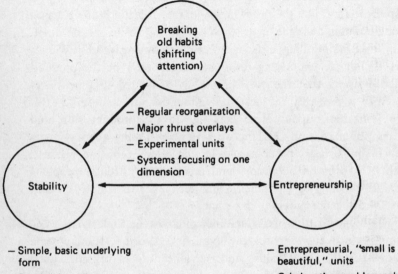

tion. (Huge, centralized staffs are primarily useful for huge coordinating exercises.) Divisions would have within their walls virtually all of the staff support they need—for example, purchasing, transportation, personnel, finance.

The third pillar, *the "habit-breaking" pillar,* encompasses in particular a willingness to reorganize regularly, and to reorganize on a "temporary" basis to attack specific thrusts (the General Motors Project Center to lead to the downsizing effort). By reorganizing regularly, we mean: (1) a willingness regularly to "hive off" new divisions as old divisions get big and bureaucratic; (2) a willingness to shift products or product lines among divisions so as to take advantage of special management talents or the need for market realignments (3M is particularly masterful at this, and turf fights seldom ensue when a product is moved from one division to another); (3) a willingness to take the top talent and bring it together on project teams aimed at solving a few central organizational problems or at executing a central organizational thrust, always with the notion that such an accommodation is temporary; and (4) a generic willingness to reorganize and reshuffle the boxes (while maintaining the integrity of the basic, central form) as needs arise.

These "habit-breaking" structural techniques are antidotes to precisely those problems that have led to the matrix organizations. Regular reorganizing is a way to meet shifting pressures without putting in place huge permanent integrating committee devices, which, in theory, take care of all possible problems from all different dimensions. The hiving off, spawning off, and product or product-line trading are similarly ways to meet shifting pressures while maintaining the integrity of the underlying form.

These three pillars, then, represent a "theoretical" response to the issues that led to the matrix organization in the first place and to the pathologies that emerged in the matrix structure as it responded to those conditions. Taken together, they also correspond closely to the managing systems of many of the excellent companies.

12

Simultaneous
Loose-Tight Properties

Simultaneous loose-tight properties, the last of our "eight basics" of excellent management practice, is mostly a summary point. It embraces much of what has come before and emerged, to our pleasant surprise, through the process of synthesis. It is in essence the co-existence of firm central direction and maximum individual autonomy—what we have called "having one's cake and eating it too." Organizations that live by the loose-tight principle are on the one hand rigidly controlled, yet at the same time allow (indeed, insist on) autonomy, entrepreneurship, and innovation from the rank and file. They do this literally through "faith"—through value systems, which our colleagues Phillips and Kennedy have suggested most managers avoid like the plague. They do it also through painstaking attention to detail, to getting the "itty-bitty, teeny-tiny things" right, as Alabama's inimitable football coach, Bear Bryant, stresses.

Loose-tight? Most businessmen's eyes glaze over when the talk turns to value systems, culture, and the like. Yet ours light up: we recall ex-chairman Bill Blackie of Caterpillar talking about Cat's commitment to "Forty-eight-hour parts service anywhere in the world." We are drawn back to a minus 60° chill factor day in Minneapolis–St. Paul, where 3M's Tait Elder talked to us about the "irrational champions" running around 3M. And we see Rene McPherson speaking to a class at Stanford. He is animated. The class asks him for the magic prescriptions with which he mastered productivity problems at Dana. He sticks his hands out in front of him, palms upright, and says, "You just keep pushing. You just keep pushing. I made every mistake that could be made. But I just kept pushing." You suspect he is serious: that really *is* all there was to it.

You think of Tom Watson, Sr., coming in after a hard day of selling pianos to farmers, and reporting to his headquarters in Painted Post, New York. And you think of what he became and why. You picture J. Willard Marriott, Sr., at that first food stand in Washington, D.C. And you see him now, at eighty-two, still worrying about a single lobby's cleanliness, although his food stand is a $2 billion enterprise. You picture Eddie Carlson working as a page boy at a Western International Hotel, the Benjamin Franklin in 1929, and marvel at the legend he has become.

Carlson doesn't blush when he talks about values. Neither did Watson—he said that values are really all there is. They lived by their values, these men—Marriott, Ray Kroc, Bill Hewlett and Dave Packard, Levi Strauss, James Cash Penney, Robert Wood Johnson. And they meticulously applied them within their organizations. They *believed* in the customer. They *believed* in granting autonomy, room to perform. They *believed* in open doors, in quality. But they were stern disciplinarians, every one. They gave plenty of rope, but they accepted the chance that some of their minions would hang themselves. Loose-tight is about rope. Yet in the last analysis, it's really about culture. Now, culture is the "softest" stuff around. Who trusts its leading analysts—anthropologists and sociologists—after all? Businessmen surely don't. Yet culture is the hardest stuff around, as well. Violate the lofty phrase, "IBM Means Service," and you are out of a job, the company's job security program to the contrary notwithstanding. Digital is crazy (soft). Digital is anarchic (soft). "People at Digital don't know who they work for," says a colleague. But they do know quality: the products they turn out work (hard). So "Soft is hard."

Patrick Haggerty says the only reason that OST (hard) works at Texas Instruments is because of TI's "innovative culture" (soft). Lew Lehr, 3M's chairman, goes around telling tales of people who have failed monumentally—but gone on, after decades of trying, to become vice presidents of the company. He's describing the loose-tight, soft-hard properties of the 3M culture.

We have talked about lots of soft traits, lots of loose traits. We have mentioned clubby, campus-like environments, flexible organizational structures (hiving off new divisions, temporary habit-breaking devices, regular reorganizations), volunteers, zealous champions, maximized autonomy for individuals, teams and divisions, regular and extensive experimentation, feedback emphasizing the positive, and strong social networks. All of these traits focus on the positive, the excitement of trying things out in a slightly disorderly (loose) fashion.

But at the same time, a remarkably tight—culturally driven/controlled—set of properties marks the excellent companies. Most have rigidly shared values. The action focus, including experimentation itself, emphasizes extremely regular communication and very quick feedback; nothing gets very far out of line. Concise paperwork (P&G's one-page memo) and the focus on realism are yet other, nonaversive ways of exerting extremely tight control. If you have only three numbers to live by, you may be sure they are all well checked out. A predominant discipline or two is in itself another crucial measure of tightness. The fact that the vast majority of the management group at 3M consists of chemical engineers, at Fluor of mechanical engineers, is another vital assurance of realism, a form of tight control.

Intriguingly, the focus on the outside, the external perspective, the attention to the customer, is one of the tightest properties of all. In the excellent companies, it is perhaps the most stringent means of self-discipline. If one is really paying attention to what the customer is saying, being blown in the wind by the customer's demands, one may be sure he is sailing a tight ship. And then there is the peer pressure: weekly Rallies at Tupperware, Dana's twice-annual Hell Weeks. Although this is not control via massive forms and incalculable numbers of variables, it is the toughest control of all. As McPherson said, it's easy to fool the boss, but you can't fool your peers. These are the apparent contradictions that turn out in practice not to be contradictions at all.

Take the quality versus cost trade-off, for example, or small versus big (i.e., effectiveness versus efficiency). They turn out in the excellent companies not to be trade-offs at all. There is a story about a GM foundry manager who led a remarkable economic turnaround; he painted the grimy interior of his foundry white, insisting that he would pay attention to quality (and housekeeping, safety), and that cost would follow. As he pointed out: "To begin with, if you are making it with good quality, you don't have to make everything twice." There is nothing like quality. It is the most important word used in these companies. Quality leads to a focus on innovativeness—to doing the best one can for every customer on every product; hence it is a goad to productivity, automatic excitement, an external focus. The drive to make "the best" affects virtually every function of the organization.

In the same way, the efficiency/effectiveness contradiction dissolves into thin air. Things of quality are produced by craftsmen, generally requiring small-scale enterprise, we are told. Activities that achieve cost efficiencies, on the other hand, are reputedly best done in large facilities, to achieve economies of scale. Except that that is not the way it works in the excellent companies. In the excellent companies, small *in almost every case* is beautiful. The small facility turns out to be the most efficient; its turned-on, motivated, highly productive worker, in communication (and competition) with his peers, outproduces the worker in the big facilities time and again. It holds for plants, for project teams, for divisions—for the entire company. So we find that in this most vital area, there really is no conflict. Small, quality, excitement; autonomy—and efficiency—are all words that belong on the same side of the coin. Cost and efficiency, over the long run, *follow* from the emphasis on quality, service, innovativeness, result sharing, participation, excitement, and an external problem-solving focus that is tailored to the customer. The revenue line does come first. But once the ball gets rolling, cost control and innovation effectiveness become fully achievable, parallel goals.

Surprisingly, the execution versus autonomy contradiction becomes a paradox, too. Indeed, one can appreciate this paradox almost anywhere. Studies in the classroom, for example, suggest that effective classes are the ones in which discipline is sure: students are expected to come to class on time; homework is regularly turned in and graded. On the other hand, those same classrooms as a general rule emphasize positive feedback, posting good reports, praise, and coaching by the teacher. Similarly, when we look at McDonald's or virtually any of the excellent companies, we find that *autonomy is a product of discipline. The discipline (a few shared values) provides the framework. It gives people confidence (to experiment, for instance) stemming from stable expectations about what really counts.*

Thus a set of shared values and rules about discipline, details, and execution can provide the framework in which practical autonomy takes place routinely. Regular experimentation takes place at 3M in a large measure because of all the tight things that surround it—extraordinarily regular communication (nothing gets far out of line), the shared values that result from the common denominator of the engineering degree, the consensus on customer problem solving that comes from a top management virtually all of whom started as down-the-line salesmen.

3M is, indeed, the tightest organization we have seen, tighter by far, in our opinion, than ITT under Geneen. At ITT, there were countless rules and variables to be measured and filed. But the dominant theme there was gamesmanship—beating the system, pulling end runs, joining together with other line officers to avoid the infamous staff "flying squads." Too much overbearing discipline of the wrong kind will kill autonomy. But the more rigid discipline, the discipline based on a small number of shared values that marks a 3M, an HP, a J&J, or a McDonald's, in fact, induces practical autonomy and experimentation throughout the organization and beyond.

The nature of the rules is crucial here. The "rules" in the excellent companies have a positive cast. They deal with quality, service, innovation, and experimentation. Their focus is on building, expanding, the opposite of restraining; whereas most companies concentrate on controlling, limiting, constraint. We don't seem to understand that rules can reinforce positive traits as well as discourage negative ones, and that the former kind are far more effective.

Even the external versus internal contradiction is resolved in the excellent companies. Quite simply, these companies are simultaneously externally focused and internally focused—externally in that they are truly driven by their desire to provide service, quality, and innovative problem solving in support of their customers; internally in that quality control, for example, is put on the back of the individual line worker, not primarily in the lap of the quality control department. Service standards likewise are substantially self-monitored. The organization thrives on internal competition. And it thrives on intense communication, on the family feeling, on open door policies, on informality, on fluidity and flexibility, on nonpolitical shifts of resources. This constitutes the crucial internal focus: the focus on people.

The skill with which the excellent companies develop their people recalls that grim conflict we first mentioned in Chapter 3: our basic need for security versus the need to stick out, the "essential tension" that the psychoanalyst Ernest Becker described. Once again the paradox, as it is dealt with in the excellent companies, holds. By offering meaning as well as money, they give their employees a mission as well as a sense of feeling great. Every man becomes a pioneer, an experimenter, a leader. The institution provides guiding belief and creates a sense of excitement, a sense of being a part of the best, a sense of producing something of quality that is generally valued. And in this way it draws out the best—from Ken Ohmae's "worker at the frontier" as from Kyoto Ceramic chairman Kazuo Inamori's "fifty percent man." The *average* worker in these companies is expected to contribute, to add ideas, to innovate in service to the customer and in producing quality products. In short, each individual—like the 9,000 leaders of PIP teams at Texas Instruments—is expected to stand out and contribute, to be distinctive. At the same time he is part of something great: Caterpillar, IBM, 3M, Disney Productions.

Finally, the last of our paradoxes involves the short-term versus long-term "trade-off." Again, we found there was no conflict at all.

We found that the excellent companies are not really "long-term thinkers." They don't have better five-year plans. Indeed, the formal plans at the excellent companies are often marked by little detail, or don't exist at all (recall the complete absence of corporate level planners in many of them).

But there is a value set—and it is a value set for all seasons. (Remember the content areas: quality, innovativeness, informality, customer service, people.) However, it is executed by attention to mundane, nitty-gritty details. Every minute, every hour, every day is an opportunity to act in support of overarching themes.

We will conclude with one strange contradiction that may really hold. We call it the smart-dumb rule. Many of today's managers— MBA-trained and the like—may be a little bit too smart for their own good. The smart ones are the ones who shift direction all the time, based upon the latest output from the expected value equation. The ones who juggle hundred-variable models with facility; the ones who design complicated incentive systems; the ones who wire up matrix structures. The ones who have 200-page strategic plans and 500-page market requirement documents that are but step one in product development exercises.

Our "dumber" friends are different. They just don't understand why every product can't be of the highest quality. They just don't understand why every customer can't get personalized service, even in the potato chip business. They are personally affronted (remember the Heineken story) when a bottle of beer goes sour. They can't understand why a regular flow of new products isn't possible, or why a worker can't contribute a suggestion every couple of weeks. Simple-minded fellows, really; simplistic even. Yes, simplistic has a negative connotation. But the people who lead the excellent companies *are* a bit simplistic. They are seemingly unjustified in what they believe the worker is capable of doing. They are seemingly unjustified in believing that every product can be of the highest quality. They are seemingly unjustified in believing that service can be maintained at a high standard for virtually every customer, whether in Missoula, Montana, or Manhattan. They are seemingly unjustified in believing that virtually every worker can contribute suggestions regularly. It is simplistic. But it may be the true key to inducing astonishing contributions from tens of thousands of people.

Of course, what one is simplistic about is vitally important. It's a focus on the external, on service, on quality, on people, on informality, those value content words we noted. And those may very well be things—the only things—worth being simplistic about. Remember the executive James Brian Quinn interviewed: he said that it was important for his people to want to be "the best" at something. He doesn't really care very much what.

But so many can't see it. There are always practical, justifiable, inevitable, sensible, and sane reasons to compromise on any of these variables. Only those simplistic people—like Watson, Hewlett, Packard, Kroc, Mars, Olsen, McPherson, Marriott, Procter, Gamble, Johnson—stayed simplistic. And their companies have remained remarkably successful.

Notes

INTRODUCTION

Page

xxi   "Society . . . is a vehicle": Ernest Becker, *Escape from Evil* (New York: Free Press, 1975), pp. 3–6, 51; and *The Denial of Death* (New York: Free Press, 1973), pp. 3–4.

xxi   An experiment in psychology: Herbert M. Lefcourt, *Locus of Control: Current Trends in Theory and Research* (Hillsdale, N.J.: Lawrence Erlbaum Associates, 1976), pp. 3–6.

CHAPTER 1: SUCCESSFUL AMERICAN COMPANIES

Page

4   Way back in 1962: Alfred D. Chandler, Jr., *Strategy and Structure: Chapters in the History of the American Industrial Enterprise* (Cambridge, Mass.: MIT Press, 1962).

5   On the shop floors: F. J. Roethlisberger and William J. Dickson, *Management and the Worker* (Cambridge, Mass.: Harvard University Press, 1939).

6   He described good managers: Chester I. Barnard, *The Functions of the Executive* (Cambridge, Mass.: Harvard University Press, 1968), chap. 5.

7   March goes even further: James G. March and Johan P. Olsen, *Ambiguity and Choice in Organizations* (Bergen, Norway: Universitetsforlaget, 1976), p. 26.

7   "He'll sit here": Richard E. Neustadt, *Presidential Power: The Politics of Leadership* (New York: Wiley, 1960), p. 9.

7   The researcher Henry Mintzberg: Henry Mintzberg, *The Nature of Managerial Work* (New York: Harper & Row, 1973), pp. 31–35.

Page

7   Pettigrew, a British researcher: Andrew M. Pettigrew, *The Politics of Organizational Decision Making* (London: Tavistock, 1973).

9   "I think an inflexible organization": William F. Dowling and Fletcher Byrom, "Conversation with Fletcher Byrom," *Organizational Dynamics,* summer 1978, p. 44.

11   *The Art of Japanese Management:* Richard Tanner Pascale and Anthony G. Athos, *The Art of Japanese Management* (New York: Simon & Schuster, 1981).

11   Footnote: "Leavitt's Diamond": Harold J. Leavitt, *Managerial Psychology,* 4th ed. (Chicago: University of Chicago Press, 1978), pp. 282ff.

14   "It's a shame": Robert L. Shook, *Ten Greatest Salespersons: What They Say About Selling* (New York: Harper & Row, 1980), p. 68.

14   "so intent on innovation": Lee Smith, "The Lures and Limits of Innovation: 3M," *Fortune,* Oct. 20, 1980, p. 84.

14   "Make sure you generate": Dowling and Byrom, p. 43.

14   "IBM's philosophy": Thomas J. Watson, Jr., *A Business and Its Beliefs: The Ideas That Helped Build IBM* (New York: McGraw-Hill, 1963), p. 13.

15   "seen as a source of ideas": Mark Shepherd, Jr., and J. Fred Bucy, "Innovation at Texas Instruments," *Computer,* September 1979, p. 84.

15   "The basic philosophy": Watson, p. 5.

15   "Never acquire a business": "The Ten Best-Managed Companies," *Dun's Review,* December 1970, p. 30.

15   "This company has never left": "P&G's New New-Product Onslaught," *Business Week,* Oct. 1, 1979, p. 79.

16   "The brainwashed members": C. Barron, "British 3M's Multiple Management," *Management Today,* March 1977, p. 56.

CHAPTER 2: THE RATIONAL MODEL

Page

31   "those who implement the plans": Mariann Jelinek, *Institutionalizing Innovation: A Study of Organizational Learning Systems* (New York: Praeger), p. 124.

32   But, as the researcher John Child: John Child, *Organization: A Guide to Problems and Practices* (New York: Harper & Row, 1977), pp. 222–23.

32   A researcher concluded recently: Stuart S. Blume, "A Managerial View of Research" (review of *Scientific Productivity,* ed. Frank M. Andrews), *Science,* Jan. 4, 1980, pp. 48–49.

Page

33   "the secular rationalist mythology": George Gilder, *Wealth and Poverty* (New York: Basic Books, 1981), p. 264.

33   "These American invaders": Steve Lohr, "Overhauling America's Business Management," *New York Times Magazine,* Jan. 4, 1981, p. 15.

33   "Nor have [the United States'] competitors": Lester C. Thurow, *The Zero-Sum Society: Distribution and the Possibilities for Economic Change* (New York: Basic Books, 1980), pp. 7–8.

34   "How quickly things change": Lohr, p. 15.

34   "The amount of money": Louis Kraar, "Japan's Automakers Shift Strategies," *Fortune,* Aug. 11, 1980, p. 109.

35   Only a few weeks later: Robert Ball, "Europe Outgrows Management American Style," *Fortune,* Oct. 20, 1980, pp. 147–48.

35   "We have created a monster": "Don't Blame the System, Blame the Managers," *Dun's Review,* September 1980, p. 88.

35   "a widely held view": Lohr, p. 58.

35   "[They] lack liberal arts literacy": Michael M. Thomas, "Businessmen's Shortcomings," *New York Times,* Aug. 21, 1980, p. D2.

35   "People with degrees": Bro Uttal, "The Animals of Silicon Valley," *Fortune,* Jan. 12, 1981, p. 94.

36   "The system is producing": *Dun's Review,* September 1980, p. 82.

36   "lacking a gut feeling": "Revitalizing the U.S. Economy," *Business Week,* June 30, 1980, p. 78.

37   "No longer does the typical career": Robert H. Hayes and William J. Abernathy, "Managing Our Way to Economic Decline," *Harvard Business Review,* July–August 1980, p. 74.

37   "You don't have much of the spirit": Lohr, p. 43.

37   "Managers don't love the product": Charles R. Day, Jr., and Perry Pascarella, "Righting the Productivity Balance," *Industry Week,* Sept. 29, 1980, p. 55.

37   "The Japanese deserve credit": Charles G. Burck, "A Comeback Decade for the American Car," *Fortune,* June 2, 1980, p. 63.

37   "While at work": Robert M. Pirsig, *Zen and the Art of Motorcycle Maintenance: An Inquiry into Values* (New York: Morrow, 1974), pp. 34–35.

38   "...the Japanese seem to have": Norman Gall, "It's Later Than We Think" (interview with William J. Abernathy), *Forbes,* Feb. 2, 1981, p. 65.

38   "American managers": Lohr, p. 23.

39   "Japanese management": Kenichi Ohmae, "Myths and Realities of Japanese Corporations" (draft), p. 11. Published as "The Myth and Reality of the Japanese Corporation," *Chief Executive,* summer 1981.

Page

40   "A lot of companies overdo it": *Dun's Review*, September 1980, p. 84.

40   "As a regimen": Dowling and Byrom, p. 40.

40   "Significantly, neither . . .": *Business Week*, June 30, 1980, p. 93.

40   "The majority of businessmen": David Ogilvy, "The Creative Chef," in *The Creative Organization*, ed. Gary A. Steiner (Chicago: University of Chicago Press, 1965), p. 206.

40   "Modelers build": Theodore Levitt, "A Heretical View of Management Science," *Fortune*, Dec. 18, 1978, p. 50.

40   "The guys were bright": "When a New Product Strategy Wasn't Enough," *Business Week*, Feb. 18, 1980, p. 143.

42   *The Structure of Scientific Revolutions:* Thomas Kuhn, *The Structure of Scientific Revolutions*, 2d ed. (Chicago: University of Chicago Press, 1970).

44   "If quantitative precision is demanded": John D. Steinbruner, *The Cybernetic Theory of Decision: New Dimensions of Political Analysis* (Princeton, N.J.: Princeton University Press, 1974), p. 328.

45   "Utterly abstract in his view": Thomas O'Hanlon, "A Rejuvenated Litton Is Once Again Off to the Races," *Fortune*, Oct. 8, 1979, p. 160.

46   "The magi inevitably talk": Lewis H. Lapham, "Gifts of the Magi," *Harper's*, February 1981, p. 11.

46   "The Mexican Sierra": John Steinbeck, *The Log from the Sea of Cortez* (New York: Viking, 1941). Quoted in Karl Weick, *Social Psychology of Organizing*, 2d ed. (Reading, Mass., Addison-Wesley, 1979), p. 29.

46   "Professional management today": Peter F. Drucker, *The Age of Discontinuity: Guidelines to Our Changing Society* (New York: Harper & Row, 1969), pp. 56–57.

47   "It is inherently easier": Steinbruner, p. 333.

47   "You think Presidents": Ibid., p. 332.

47   "The financial people": "Mobil's Successful Exploration," *Business Week*, Oct. 13, 1980, p. 114.

47   "We believe that during the past two decades": Hayes and Abernathy, pp. 70–71.

47   "Creative thought": Gilder, p. 262.

47   "When they were built": Ibid., p. 252.

48   "[There is a] rockbound difference": Robert K. Merton, *Social Theory and Social Structure*, enlarged ed. (New York: Free Press, 1968), p. 4.

48   "It is no use looking": Horace F. Judson, *Search for Solutions* (New York: Holt, Rinehart and Winston, 1980), p. 3.

Page
48   "The quickest way to understand": Alexander Cockburn, James Ridgeway and Andrew Cockburn, "The Pentagon Spends Its Way to Impotence," *Village Voice,* Feb. 18, 1981, p. 11.
49   "Why are these new administrative structures": Chris Argyris, "Today's Problems with Tomorrow's Organizations," *Journal of Management Studies,* February 1967, pp. 34–40.
50   "Of all the things": Fletcher Byrom, speech delivered to Carnegie-Mellon GSIA, 1976.
50   "You've got to avoid": "Lessons of Leadership: David Packard," *Nation's Business,* January 1974, p. 42.
50   "Most Japanese companies": Ohmae, pp. 5, 20.
51   Product after product: Jelinek, p. 54.

CHAPTER 3: MAN WAITING FOR MOTIVATION

Page
56   In a recent psychological study: David G. Myers, *The Inflated Self.* Mentioned in "How Do I Love Me? Let Me Count the Ways," *Psychology Today,* May 1980, p. 16.
58   "attribution theory": Lee Ross, "The Intuitive Psychologist and His Shortcomings," in *Advances in Experimental Social Psychology,* vol. 10, ed. Leonard Berkowitz (New York: Academic Press, 1977), pp. 173–220.
58   In one experiment, adults: Russell A. Jones, *Self-Fulfilling Prophecies: Social, Psychological and Physiological Effects of Expectancies* (Hillsdale, N.J.: Lawrence Erlbaum Associates, 1977), p. 167.
59   "In a study of school teachers": Warren Bennis, *The Unconscious Conspiracy: Why Leaders Can't Lead* (New York: AMACOM, 1976), p. 174.
59   "[our] behavior continues": Arthur Koestler, *The Ghost in the Machine* (New York: Macmillan, 1967), p. 274.
59   "the psychoanalytic emphasis": Ernest Becker, *The Denial of Death* (New York: Free Press, 1973), p. 94.
60   "One fact recurs": Henry Mintzberg, "Planning on the Left Side and Managing on the Right," *Harvard Business Review,* July–August 1976, p. 53.
61   "It's so beautiful": "How to Get a Bright Idea," *The Economist,* Dec. 27, 1980, p. 61.
61   "When you have something simple": Horace F. Judson, *Search for Solutions* (New York: Holt, Rinehart and Winston, 1980), p. 22.
61   Two experimental psychologists: Amos Tversky and Daniel Kahneman, "Judgment Under Uncertainty: Heuristics and Biases," *Science,* Sept. 27, 1974, p. 1124.

Page

62   "There's a story": Gregory Bateson, *Mind and Nature: A Necessary Unity* (New York: Bantam Books, 1980), p. 14.

66   The something else, Simon believes: H. A. Simon, "Information Processing Models of Cognition," *Annual Review of Psychology,* vol. 30 (Palo Alto, Calif.: Annual Reviews, 1979), p. 363.

67   "technology of behavior": B. F. Skinner, *Beyond Freedom and Dignity* (New York: Knopf, 1971), p. 5.

68   "The person who has been punished": Ibid., p. 81.

70   As Skinner notes: Ibid., pp. 34ff.

70   At Foxboro, a technical advance: Allan A. Kennedy, personal communication.

71   "social comparison theory": Leon Festinger, "A Theory of Social Comparison Processes," *Human Relations* 7 (1954): 117–40.

72   Edward Deci of the University of Rochester: Edward L. Deci, "The Effects of Contingent and Non-Contingent Rewards and Controls on Intrinsic Motivations," *Organizational Behavior and Human Performance* 8 (1972): 217–29.

73   "You more likely act": Jerome S. Bruner, *On Knowing: Essays for the Left Hand* (New York: Atheneum, 1973), p. 24.

73   "LaPiere, a white professor": Jonathan L. Freedman, David O. Sears, and J. Merrill Carlsmith, *Social Psychology,* 3d ed. (Englewood Cliffs, N.J.: Prentice-Hall, 1978), p. 299.

74   "foot-in-the-door research": Jonathan L. Freedman and Scott C. Fraser, "Compliance Without Pressure: The Foot-in-the-Door Technique," *Journal of Personality and Social Psychology* 4 (1966): 195–202.

74   He lists major leadership tasks: James Brian Quinn, "Formulating Strategy One Step at a Time," *Journal of Business Strategy,* winter 1981, pp. 57–59.

75   "The world is an illusion": Robert L. Forward, "Spinning New Realities," *Science 80,* December 1980, p. 40.

75   "If we hope to live": Bruno Bettelheim, *On the Uses of Enchantment: The Meaning and Importance of Fairy Tales* (New York: Knopf, 1976), p. 3.

76   "They speak of things": Oscar Shisgall, *Eyes on Tomorrow: The Evolution of Procter & Gamble* (Chicago: J. G. Ferguson, 1981), p. xi.

76   "he who has a *why* to live for": Viktor E. Frankl, *Man's Search for Meaning* (New York: Pocket Books, 1963), p. 164.

76   "Man is a stubborn seeker": John W. Gardner, *Morale* (New York: Norton, 1978), p. 15.

78   Stanley Milgram's experiments: Stanley Milgram, *Obedience to Authority: An Experimental View* (New York: Harper & Row, 1974).

Page
79 a "prison" experiment: Philip Zimbardo and Greg White, "The Stanford Prison Experiment: A Simulation of the Study of the Psychology of Imprisonment Conducted August 1971 at Stanford University" (script for slide show), n.d.

80 "Man thus has the absolute tension": Becker, *Denial of Death*, pp. 153–54.

80 The typical experiment here: Jones, p. 133.

81 A subject allowed to dip: Gerald R. Salancik, "Commitment and the Control of Organizational Behavior and Belief," in *New Directions in Organizational Behavior*, ed. Barry M. Staw and Gerald R. Salancik (Chicago: St. Clair Press, 1977), pp. 20ff.

82 But Burns has posited: James MacGregor Burns, *Leadership* (New York: Harper & Row, 1978).

83 "This absolutely central value": Ibid., pp. 13, 18–19.

83 "[Transforming leadership] occurs": Ibid., p. 20.

83 "The fundamental process": Ibid., p. 40.

84 "His true genius": Ibid., p. 254.

84 "Managers prefer working": Abraham Zaleznick, "Managers and Leaders: Are They Different?" *Harvard Business Review*, May–June 1977, p. 72.

84 "[We] set out to find": David C. McClelland, *Power: The Inner Experience* (New York: Irvington, 1975), pp. 259–60.

84 "You have to believe": Ray Kennedy, "Howard Head Says, 'I'm Giving Up the Thing World,' " *Sports Illustrated*, Sept. 29, 1980, p. 72.

84 "We have slowly discovered": James B. Quinn, "Strategic Goals: Process and Politics," *Sloan Management Review*, fall 1977, p. 26.

85 the leader as "social architect": Bennis, p. 165.

85 "The inbuilding of purpose": Philip Selznick, *Leadership in Administration: A Sociological Interpretation* (New York: Harper & Row, 1957), pp. 17, 28, 149–50, 152–53.

85 "be true to our own aesthetic": Jill Gerston, "Tiffany's Unabashed Guardian of Good Taste Relinquishes Helm," *San Francisco Examiner*, Jan. 5, 1981, p. C2.

85 "beauty in a hamburger bun": Ray Kroc, *Grinding It Out: The Making of McDonald's* (New York: Berkley, 1977), p. 98.

CHAPTER 4: MANAGING AMBIGUITY AND PARADOX

Page
89 "The test of a first-rate intelligence": F. Scott Fitzgerald, "The Crack-up," in *American Literary Masters*, vol. 2, ed. Charles R. Anderson (New York: Holt, Rinehart and Winston, 1965), p. 1007.

Page
90   William Manchester, in *Good-bye, Darkness:* William Manchester, *Good-bye, Darkness: A Memoir of the Pacific War* (Boston: Little, Brown, 1980), pp. 233–37.

91   Richard Scott of Stanford: W. Richard Scott, "Theoretical Perspectives," in *Environments and Organizations,* by Marshall W. Meyer and Associates (San Francisco: Jossey-Bass, 1978).

94   "This volume is an attempt": Douglas McGregor, *The Human Side of Enterprise* (New York: McGraw-Hill, 1960), pp. vi, vii.

94   "If there is a single assumption": Ibid., p. 18.

95   "the assumption of the mediocrity": Ibid., p. 34.

95   "(1) that the average human": Ibid., pp. 49–50.

95   "but is in fact a theory": Ibid., p. 35.

95   "(1) that the expenditure": Ibid., pp. 47–48.

96   "The assumptions of Theory Y": Ibid., p. 56.

97   "Barnard's aim": Kenneth R. Andrews, "Introduction to the Anniversary Edition," in *The Functions of the Executive,* by Chester I. Barnard (Cambridge, Mass.: Harvard University Press, 1968), p. vii.

97   "The essential functions": Chester I. Barnard, *The Functions of the Executive* (Cambridge, Mass.: Harvard University Press, 1968), p. 217.

97   "It has already been made clear": Ibid., p. 231.

98   "The common sense of the whole": Ibid., pp. 238–39.

98   "The term 'organization'": Selznick, pp. 5ff., 40, 135–36.

100 *Organization and Environment:* Paul R. Lawrence and Jay W. Lorsch, *Organization and Environment: Managing Differentiation and Integration* (Homewood, Ill.: Richard D. Irwin, 1967).

101 "Organizations have staff": Karl E. Weick, *The Social Psychology of Organizing,* 2d ed. (Reading, Mass.: Addison-Wesley, 1979), p. 49.

101 "It forces people": Ibid., p. 50.

101 The new metaphors: Ibid., p. 47.

101 "Diverse as they are": Ibid.

101 March and Simon's joint work: James G. March and Herbert A. Simon, *Organizations* (New York: Wiley, 1958).

102 *Ambiguity and Choice in Organizations:* March and Olsen, *Ambiguity and Choice in Organizations.*

102 "This book is about": Weick, p. 1.

103 "What Delta has going": "The Five Best-Managed Companies," *Dun's Review,* December 1977, p. 60.

103 "OST would be sterile": Mark Shepherd, Jr., and J. Fred Bucy, "Innovation at Texas Instruments," *Computer,* September 1979, p. 89.

Page

103   "The reliability of Maytag": Edmund Faltermayer, "The Man Who Keeps Those Maytag Repairmen Lonely," *Fortune,* November 1977, p. 192.

103   "Firms operating out of Rochester": Stanley M. Davis, "Establishing a New Context for Strategy, Organization and Executive Pay," in *Executive Compensation in the 1980s,* ed. David J. McLaughlin (San Francisco: Pentacle Press, 1980), p. 29.

104   "dominating business idea": Richard Normann, *Management and Statesmanship* (Stockholm: Scandinavian Institutes for Administrative Research, 1976), p. 275.

104   "missionary configuration": Henry Mintzberg, *The Structuring of Organizations: A Synthesis of the Research* (Englewood Cliffs, N.J.: Prentice-Hall, 1979), p. 480.

104   "The [leader] not only creates": Andrew M. Pettigrew, "The Creation of Organisational Cultures" (paper presented to the Joint EIASM–Dansk Management Center Research Seminar, Copenhagen, May 18, 1976), p. 11.

104   "systems composed of ideas": Joanne Martin, "Stories and Scripts in Organizational Sellings," Research Report no. 543 (rev.) (Graduate School of Business, Stanford University, July 1980), p. 3.

105   "It is not so much the articulation": Bennis, p. 93.

105   *Business Week* legitimated: "Corporate Culture: The Hard-to-Change Values That Spell Success or Failure," *Business Week,* Oct. 27, 1980, pp. 148–60.

105   *The Organization Man:* William H. Whyte, Jr., *The Organization Man* (New York: Simon & Schuster, 1956).

105   "The company gives me": Steven Rothman, "More than Money," *D & B Reports,* March–April, p. 12.

107   "[we] need to supplement": James G. March, "The Technology of Foolishness," in *Readings in Managerial Psychology,* 3d ed., ed. Harold J. Leavitt, Louis R. Pondy, and David M. Boje (Chicago: University of Chicago Press, 1980), p. 576.

107   "Rather than an analyst": James G. March, "Footnotes to Organizational Change" (unpublished manuscript, n.d.), p. 20.

107   "such a vision": Ibid., p. 35.

107   "organizational design": Ibid., p. 22.

107   "loosely coupled systems": Karl E. Weick, "Educational Organizations as Loosely Coupled Systems," *Administrative Science Quarterly* 21 (1976): 1–19.

107   "The more one delves": Weick, p. 120.

107   "Unjustified variation": Ibid., p. 193.

108   "retrospective sense making": Ibid., p. 202.

Page

108   "an impoverished, shallow environment": Ibid., p. 193.

108   "No one is ever free": Ibid.

108   "If you place in a bottle": Karl E. Weick, "The Management of Organizational Change Among Loosely Coupled Elements" (unpublished manuscript, December 1981), pp. 3–4.

108   "This episode speaks": Ibid., p. 4.

109   "Marketing Myopia"; Theodore Levitt, "Marketing Myopia," *Harvard Business Review,* July–August 1960.

110   "Assuming that an industry": Burton H. Klein, *Dynamic Economics* (Cambridge, Mass.: Harvard University Press, 1977), p. 17.

110   "The very process": Gilder, p. 79.

110   IBM's real management magic: Robert Sobel, *IBM: Colossus in Transition* (New York: Times Books, 1981), p. 346.

111   "Businessmen will have to learn": Drucker, p. 54.

111   "constant reorganization": Norman Macrae, "The Coming Entrepreneurial Revolution: A Survey," *The Economist,* Dec. 25, 1976, pp. 41, 43.

111   ". . . we can predict": H. Igor Ansoff, "Corporate Structure Present and Future," Vanderbilt University Working Paper 74-4, February 1974, p. 17.

112   "Ten years ago": "It Seemed Like a Good Idea at the Time," *Science 82,* January/February 1982, p. 86.

113   *Markets and Hierarchies:* Oliver E. Williamson, *Markets and Hierarchies: Analysis and Antitrust Implications* (New York: Free Press, 1975).

115   Stephen Jay Gould: Stephen Jay Gould, *The Panda's Thumb: More Reflections in Natural History* (New York: Norton, 1980), p. 51.

115   IBM product introduction: James Brian Quinn, "Technological Innovation, Entrepreneurship, and Strategy," *Sloan Management Review,* spring 1979, p. 25.

116   computers were seen to have: James M. Utterback, "Patterns of Industrial Innovation," in *Technology, Innovation, and Corporate Strategy: A Special Executive Seminar Presented by the Massachusetts Institute of Technology, November 17, 1978* (Cambridge, Mass.: Industrial Liaison Program, MIT, 1978).

116   *The External Control of Organizations:* Jeffrey Pfeffer and Gerald R. Salancik, *The External Control of Organizations: A Resource Dependence Perspective* (New York: Harper & Row, 1978).

116   "The central thesis": Ibid., p. xi.

117   "It implies special connections": Utterback, pp. 37–38.

CHAPTER 5: A BIAS FOR ACTION

Page

121   Both Warren Bennis: Warren Bennis, "The Temporary Society," in *The Temporary Society* by Warren G. Bennis and Philip E. Slater (New York: Harper & Row, 1968).

121   and Alvin Toffler: Alvin Toffler, *The Third Wave* (New York: Morrow, 1980).

122   "Visible management": Richard T. Pascale, "The Role of the Chief Executive in the Implementation of Corporate Policy: A Conceptual Framework," Research Paper no. 357 (Graduate School of Business, Stanford University, February 1977), pp. 37, 39.

122   HP treats MBWA: William R. Hewlett and David Packard, *The HP Way* (Palo Alto, Calif.: Hewlett-Packard, 1980), p. 10.

122   Corning Glass installed: Edward Meadows, "How Three Companies Increased Their Productivity," *Fortune,* Mar. 10, 1980, p. 95.

124   "The Chase senior executive's": Alena Wels, "How Citicorp Restructured for the Eighties," *Euromoney,* April 1980, p. 13.

124   "You've got to keep coming up": Susan Benner, "He Gave Key People a Reason to Stay with the Company," *Inc.,* September 1980, p. 46.

125   The prime ingredient: Robert J. Flaherty, "Harris Corp.'s Remarkable Metamorphosis," *Forbes,* May 26, 1980, p. 46.

127   "The essential building block": Ezra F. Vogel, *Japan as Number One: Lessons for America* (Cambridge, Mass.: Harvard University Press, 1979), pp. 143–45.

127   "a fluid, project-oriented environment": Shepherd and Bucy, "Innovation at Texas Instruments," p. 88.

131   The necessity of open communications: Frederick P. Brooks, Jr., *The Mythical Man-Month: Essays on Software Engineering* (Reading, Mass.: Addison-Wesley, 1978).

131   "Everyone had the authority": Ibid., p. 67.

133   The principal vehicle: Charles G. Burck, "How GM Turned Itself Around," *Fortune,* Jan. 16, 1978.

135   "Rather than test": R. Jeffrey Smith, "Shuttle Problems Compromise Space Program," *Science,* November 1979, pp. 910–11.

136   "They surprised themselves": Mariann Jelinek, *Institutionalizing Innovation: A Study of Organizational Learning Systems* (New York: Praeger, 1979), p. 78.

137   "Our approach is to make": Smith, "3M," p. 94.

137   "testing fetish": *Business Week,* Oct. 1, 1979, p. 80.

Page

137   "Bloomie's is the only": Mark Stevens, *"Like No Other Store in the World": The Inside Story of Bloomingdale's* (New York: Crowell, 1979), p. 138.

137   "I lean more to being": William Shockley, "A Case: Observations on the Development of the Transistor," in *The Creative Organization,* ed. Gary A. Steiner (Chicago: University of Chicago Press, 1965), pp. 139–40.

138   "The most important word": David Ogilvy, *Confessions of an Advertising Man* (New York: Atheneum, 1980), p. 86.

138   "Have you heard of zoom lenses?": Peter G. Peterson, "Some Approaches to Innovation in Industry-Discussion," in *The Creative Organization,* pp. 191–92.

139   In his classic work: S. I. Hayakawa, *Language in Thought and Action* (London: Allen & Unwin, 1974).

141   "Most favorable results": Donald D. Holt, "How Amoco Finds All That Oil," *Fortune,* Sept. 8, 1980, p. 51.

142   "It has been my observation": Harold Guetzkow, "The Creative Person in Organizations," in *The Creative Organization,* p. 49.

143   "A disc drive": Bro Uttal, "Storage Technology Goes for the Gold," *Fortune,* Apr. 6, 1981, p. 58.

146   "I had no authority": Isadore Barmash, *For the Good of the Company: Work and Interplay in a Major American Corporation* (New York: Grosset & Dunlap, 1976), pp. 43–44, 52–54.

149   The essential idea: Robert H. Schaffer, "Make Success the Building Block," *Management Review,* August 1981, pp. 47, 49–51.

149   "Select *one* branch": Ibid., p. 51.

150   "Deupree strongly disliked": Oscar Schisgall, *Eyes on Tomorrow: The Evolution of Procter & Gamble* (Chicago: J. G. Ferguson, 1981), p. 120.

151   "A brief written presentation": Thomas J. Peters, "The 1-Page Memo (and Other Draconian Measures)" (unpublished manuscript, April 1980), p. 1.

151   "They are so thorough": "P&G's New New-Product Onslaught," *Business Week,* Oct. 1, 1979, p. 80.

152   "They are a very deliberate": Lee Smith, "A Superpower Enters the Soft-Drink Wars," *Fortune,* June 30, 1980, p. 77.

152   Jorge Diaz Serrano: Alan Riding, "Mexico's Oil Man Proved His Point," *New York Times,* July 16, 1978, p. F5.

152   "I am known": "Paper Work Is Avoidable (If You Call the Shots)," *Wall Street Journal,* June 17, 1977, p. 24.

153   "Division presidents": Thomas J. Peters, "Management Systems:

Page

The Language of Organizational Character and Competence," *Organizational Dynamics,* summer 1980, p. 15.

153   "Although head office": Geoffrey Foster, "Dana's Strange Disciplines," *Management Today,* September 1976, p. 61.

153   "Three to five objectives": John W. Hanley, "Monsanto: The Management Style" (internal communication, September 1974), p. 10.

CHAPTER 6: CLOSE TO THE CUSTOMER

Page

156   "Probably the most important": Lewis H. Young, "Views on Management" (speech to Ward Howell International, Links Club, New York, Dec. 2, 1980), p. 5.

157   Although he's not a company: Pages 157–158 rely heavily upon "Joe Girard," in Shook, *Ten Greatest Salespersons,* pp. 7–24.

158   "Joe's customers": Ibid., p. 24.

159   "In time, good service": Watson, *A Business and Its Beliefs,* pp. 29, 32.

161   "getting the order": Shook, pp. 55–73.

161   "You have to remember": "No. 1's Awesome Strategy," *Business Week,* June 8, 1981, p. 86.

162   "It keeps the person involved": Ibid., p. 88.

165   The best outside analysis: Pages 165–167 rely heavily upon Dinah Nemeroff, *Service Delivery Practices and Issues in Leading Consumer Service Businesses: A Report to Participating Companies* (New York: Citibank, April 1980).

167   One of the best examples of service: Pages 167–168 rely heavily upon N. W. Pope, "Mickey Mouse Marketing," *American Banker,* July 25, 1979, and Pope, "More Mickey Mouse Marketing," *American Banker,* Sept. 12, 1979.

168   "Traditions I": Pope, "Mickey Mouse Marketing," p. 14.

169   "Nearly every operator": Victor F. Zonana, "Boeing's Sale to Delta Gives It Big Advantage Over U.S. Competitors," *Wall Street Journal,* Nov. 13, 1980, pp. 1, 20.

169   "We've tried to build": Harold Mansfield, *Vision: The Story of Boeing* (New York: Duell, Sloan & Pearce, 1966), pp. 361–62.

171   Caterpillar offers customers: "Caterpillar: Sticking to Basics to Stay Competitive," *Business Week,* May 4, 1981, p. 74.

171   "The company's operating principles": Gilbert Cross, "The Gentle Bulldozers of Peoria," *Fortune,* July 1963, p. 167.

171   "Product quality is something": "Caterpillar," *Business Week,* May 4, 1981, p. 74.

Page

172   "The Company even conducts": Ibid., p. 77.

172   "We adopted a firm policy": William L. Naumann, "The Story of
      Caterpillar Tractor Co." (speech to Newcomen Society of North
      America, Chicago, Mar. 17, 1977), p. 16.

172   "users can count": Ibid.

172   "A machine made": Ibid.

172   "If I had a brick": Kroc, *Grinding It Out,* p. 91.

173   "Quality is the first word": *McDonald's Corporation 1980 Annual
      Report* (Oak Brook, Ill., 1980), p. 4.

174   "The problem is consistency": "Burger King Looks for Consisten-
      cy," *Sun,* July 1980.

174   "growth is not our principal goal": *Digital Equipment Corporation
      1979 Annual Report* (Maynard, Mass.: Digital Equipment Corpora-
      tion, 1979), p. 3.

174   "Ten years' trouble-free operation": Edmund Faltermayer, "The
      Man Who Keeps Those Maytag Repairmen Lonely," *Fortune,* No-
      vember 1977, p. 193.

174   "Maytag built its reputation": Lawrence Ingrassia, "Staid Maytag
      Puts Its Money on Stoves but May Need to Invest Expertise, Too,"
      Wall Street Journal, July 23, 1980, p. 25.

175   "Cosmetics don't sit well": Bill Abrams, "P&G May Give Crest a
      New Look After Failing to Brush Off Rivals," *Wall Street Journal,*
      Jan. 8, 1981, p. 21.

175   *Hewlett-Packard's* Computer Systems: Pages 175–177 rely heavily
      upon Bill Hooper, Susan Konn, Robin Rakusin, Mike Sanders, and
      Tom Shannon, "The Management of Quality in the Computer Ser-
      vices Division of Hewlett-Packard Company" (unpublished manu-
      script, Graduate School of Business, Stanford University, Feb. 25,
      1982).

178   "The company is seldom first": Kathleen K. Wiegner, "The One to
      Watch," *Forbes,* Mar. 2, 1981, p. 60.

178   "While sometimes a competitor...": *Dun's Review,* December
      1978, p. 40.

178   "It was rarely the first": Catherine Harris, "What Ails IBM?" *Fi-
      nancial World,* May 15, 1981, p. 17.

178   "Caterpillar is rarely the first": *Business Week,* May 4, 1981, p. 77.

179   "Deere isn't saying": Harlan S. Byrne, "Deere & Co. Farm-
      Machinery Leadership Helps Firm Weather the Industry's Slump,"
      Wall Street Journal, Feb. 20, 1981, p. 48.

179   "I must keep telling": David B. Tinnin, "The Heady Success of Hol-
      land's Heineken," *Fortune,* Dec. 16, 1981, p. 169.

Page

180   "The customer who is looking": Treadwell Davison, personal communication (Graduate School of Business, Stanford University, February 1982).

180   "On the surface there is nothing": Alistair Mant, *The Rise and Fall of the British Manager*, rev. ed. (London: Pan Books Ltd.), pp. 108–9.

182   "Each department": Walter McQuade, "Making a Drama Out of Shopping," *Fortune*, Mar. 24, 1980, p. 107.

182   "The category had been asleep": Howard Rudnitsky and Jay Gissen, "Winning Big by Thinking Small," *Forbes*, Sept. 28, 1981, p. 106.

183   "Our organization does not believe": Lewis W. Lehr, "How 3M Develops Entrepreneurial Spirit Throughout the Organization," *Management Review*, October 1980, p. 31.

184   "the new technology enters": James M. Utterback, "Patterns of Industrial Innovations," in *Technology Innovation and Corporate Strategy: A Special Executive Seminar Presented by the Massachusetts Institute of Technology, November 17, 1978* (Cambridge, Mass.: Industrial Liaison Program, MIT, 1978), p. 3.

185   "So long as you spend": Howard Rudnitsky, "Will It Play in Toledo?" *Forbes*, Nov. 10, 1980, p. 198.

185   "Using our data better": Herbert Meyer, "How Fingerhut Beat the Recession," *Fortune*, Nov. 17, 1980, p. 103.

188   A lead *Fortune* article: Bro Uttal, "The Gentlemen and the Upstarts Meet in a Great Mini Battle," *Fortune*, Apr. 23, 1979, pp. 98–108.

190   "A month before your son": Meyer, pp. 103–4.

191   When Neiman-Marcus opened: Stanley Marcus, *Minding the Store* (New York: New American Library, 1975), p. 3.

194   P&G says it got 200,000 calls: *The Procter & Gamble Company Annual Report* (Cincinnati: Procter & Gamble, 1979), p. 13.

194   Not long ago Von Hippel: Eric A. Von Hippel, "Users as Innovators," *Technology Review*, January 1978, pp. 31–39.

194   Von Hippel reports: Ibid., pp. 32–33.

195   In 1873, for $68: Ed Cray, *Levi's* (Boston: Houghton Mifflin, 1978), pp. 21–22.

195   "They rely on customers": Uttal, p. 100.

195   "They will be more influenced": "Wang Labs Challenges the Goliaths," *Business Week*, June 4, 1979, p. 100.

196   Among the most extensive innovation studies: Pages 196–197 rely heavily upon *Success and Failure in Industrial Innovation: Report on Project SAPPHO* (Science Policy Research Unit, University of

Sussex, London: Centre for the Study of Industrial Innovation, February 1972), and Roy Rothwell, "SAPPHO Updated—Project SAPPHO, phase II," unpublished manuscript (Science Policy Research Unit, University of Sussex, July 1973).

CHAPTER 7: AUTONOMY AND ENTREPRENEURSHIP

Page

200   "The new idea": Modesto A. Maidique, "Entrepreneurs, Champions, and Technological Innovation," *Sloan Management Review,* winter 1980, p. 60.

200   "small firms produced": Lucien Rhodes and Cathryn Jakobson, "Small Companies: America's Hope for the 80s," *Inc.,* April 1981, p. 44.

200   Studying the same subject: Burton H. Klein, *Dynamic Economics* (Cambridge, Mass.: Harvard University Press, 1977).

200   Veronica Stolte-Heiskanen: Blume, "A Managerial View of Research," *Science,* Jan. 4, 1980, p. 48.

202   "He was possessed": Quinn, "Technological Innovation," p. 20.

202   "In 1946 Head went off": Kennedy, "Howard Head Says, 'I'm Giving Up the Thing World,' " pp. 68–70.

204   "Committed champions were encouraged": Quinn, p. 25.

205   "Like most companies, GE": Niles Howard and Susan Antilla, "Putting Innovation to Work," *Dun's Review,* p. 78.

206   "The trouble with much of the advice": Theodore Levitt, "Ideas Are Useless Unless Used," *Inc.,* February 1981, p. 96.

208   "One-man shows": William E. Souder, "Encouraging Entrepreneurship in the Large Corporations," *Research Management,* May 1981, p. 19.

210   "Dick Gelb says": Thomas Jaffe, "When Opportunity Knocks," *Forbes,* Oct. 13, 1980, pp. 96–100.

210   "Standard likes to drill": Donald D. Holt, "How Amoco Finds All that Oil," *Fortune,* Sept. 8, 1980, p. 51.

212   "We were trying to present": William Dowling and Edward Carlson, "Conversation with Edward Carlson," *Organizational Dynamics,* spring 1979, p. 58.

213   "To me, Schlumberger": "Schlumberger: The Star of the Oil Fields Tackles Semiconductors," *Business Week,* Feb. 16, 1981, p. 60.

213   "the salesman's company": C. Barron, "British 3M's Multiple Management," *Management Today,* March 1977, p. 57.

216   One of Roger Smith's stated priorities: Amanda Bennett, "GM's

Page

Smith Wants Leaner Firm, More Rivalry Among Its Divisions," *Wall Street Journal,* May 21, 1981, p. 43.

216   Management decided, even then: Oscar Schisgall, *Eyes on Tomorrow: The Evolution of Procter & Gamble* (Chicago: J. G. Ferguson, 1981), p. 162.

217   "Digital's idiosyncratic growth strategy": Bro Uttal, "The Gentlemen and the Upstarts," p. 101.

218   chief executive Ken Olsen: Maidique, p. 67.

219   "Proponents of successful ideas": Ibid., p. 60.

220   MIT's Thomas Allen: Thomas J. Allen, "Communications in the Research and Development Laboratory," *Technology Review,* October–November 1967.

222   "dreamers, heretics, gadflies": Advertisement in *Newsweek,* Aug. 11, 1980, p. 6.

223   GE came up with a "toy shop": Gene Bylinsky, "Those Smart Young Robots on the Production Line," *Fortune,* Dec. 17, 1979, p. 93.

223   one of J&J's tenets: Lee Smith, "J&J Comes a Long Way from Baby," *Fortune,* June 1, 1981, p. 66.

223   "You need the ability to fail": Marshall Loeb, "A Guide to Taking Charge," *Time,* Feb. 25, 1980, p. 82.

225   Among the new products: Lee Smith, "The Lures and Limits of Innovation: 3M," *Fortune,* Oct. 20, 1980, p. 84.

225   "Whenever anything is being accomplished": Peter F. Drucker, *Adventures of a Bystander* (New York: Harper & Row, 1979), p. 255.

225   "What keeps them satisfied": Smith, *Fortune,* Oct. 20, 1980, p. 86.

226   "The team members are recruited": Edward B. Roberts, "Managing New Technical Ventures," in *Technology, Innovation, and Corporate Strategy: A Special Executive Seminar* (Cambridge, Mass.: Industrial Liaison Program, MIT, 1978), pp. 121–22.

226   "They say," notes MIT's Edward Roberts: Ibid., p. 122.

227   "The individual involved": Ibid., pp. 125–26.

227   "If you want to stop a project": Ibid., p. 120.

228   "The salesmen who visited": Smith, *Fortune,* Oct. 20, 1980, p. 90.

229   "Fifteen to twenty or more": Lehr, p. 38.

229   "Because of 3M's diversity": Roberts, p. 123.

229   "If the product idea can meet": Ibid.

229   "We got into the business": Lehr, p. 31.

231   "The obsession with invention": Smith, *Fortune,* Oct. 20, 1980, p. 90.

Page
231   "The original salesmen": Ibid.
232   "Our experience": Roberts, p. 123.

CHAPTER 8: PRODUCTIVITY THROUGH PEOPLE

Page
235   The Navy . . . assumes: Elmo R. Zumwalt, Jr., *On Watch: A Memoir* (New York: Times Books, 1976), p. 183.
235   "Are these men and women": Given to authors by Gary D. Bello, Stanford Sloan Program, March 1982.
236   "What I tried hardest to do": Zumwalt, p. 186.
237   "I now knew from experience": Ibid., p. 185.
237   "Japanese management keeps telling": Kenichi Ohmae, "The Myth and Reality of the Japanese Corporation," p. 29.
237   "People will flood you": Robert Lubar, "Rediscovering the Factory," *Fortune*, July 13, 1981, p. 60.
237   "I was operations manager": Sam T. Harper, personal communication (Graduate School of Business, Stanford University, January 1982).
238   "IBM's philosophy": Watson, *A Business and Its Beliefs*, p. 13.
242   RMI: Pages 242–43 are based, in part, upon Cindy Ris, "Big Jim Is Watching at RMI Co., and Its Workers Like It Just Fine," *Wall Street Journal*, Aug. 4, 1980, p. 15.
244   "I feel that in general terms": Hewlett and Packard, *The HP Way*, p. 3.
246   Wal-Mart: Pages 246–48 rely heavily upon Lynda Schuster, "Wal-Mart Chief's Enthusiastic Approach Infects Employees, Keeps Retailer Growing," *Wall Street Journal*, Apr. 20, 1982, p. 21.
249   "Until we believe": Rene C. McPherson, "The People Principle," *Leaders*, January–March 1980, p. 52.
250   "We didn't waste time": "Rene McPherson: GSB Deanship Is His Way to Reinvest in the System," *Stanford GSB*, fall 1980–81, p. 15.
252   "The way you stay fresh": Ibid.
252   "I guess I got laid off": George H. Labovitz, Speech to the Opening Assembly, Western Hospital Association, Anaheim, Calif., Apr. 27, 1981.
253   Delta Airlines: Pages 253–255 are based, in part, upon Margaret R. Keefe Umanzio, "Delta Is Ready," unpublished manuscript (San Francisco: McKinsey & Co., July 1981).
253   "Stewardesses, for instance": Janet Guyon, " 'Family Feeling' at Delta Creates Loyal Workers, Enmity of Unions," *Wall Street Journal*, July 7, 1980, p. 13.

Page
253 "My rug has to be cleaned": "W. T. Beebe: The Gold Winner," *Financial World*, Mar. 15, 1978, p. 21.
254 "In February 1979": Guyon, p. 13.
255 "That's important": Ibid.
255 "History shows": "The Five Best-Managed Companies," *Dun's Review*, December 1977, p. 50.
255 "I believe that *less is more*": Kroc, *Grinding It Out*, p. 143.
255 "A well-run restaurant": Ibid., p. 99.
255 "I emphasize the importance": Ibid., p. 101.
256 "The book" at McDonald's: Jeremy Main, "Toward Service Without a Snarl," *Fortune*, Mar. 23, 1981, p. 66.
256 "Debbie Thompson": Ibid.
256 "The American flag": Susan Saiter Anderson, "Hamburger U. Offers a Break," *Survey of Continuing Education (New York Times)*, Aug. 30, 1981, pp. 27–28.
258 Walk into IBM's: Allan J. Mayer and Michael Ruby, "One Firm's Family," *Newsweek*, Nov. 21, 1977, p. 84.
258 "T. J. Watson didn't move in": Watson, *A Business and Its Beliefs*, p. 15.
259 "IBM produced parts": Ibid., pp. 15–16.
259 "providing showers on company premises": Ibid., p. 17.
259 "Almost every kind of fanfare": Ibid., p. 18.
259 "born homilist": Gil Burck, "International Business Machines," *Fortune*, January 1940, p. 41.
260 "Above all, we seek": Shook, *Ten Greatest Salespersons*, p. 73.
261 "If you look at the entrepreneurship": Thomas L. Friedman, "Talking Business," *New York Times*, June 9, 1981, p. D2.
263 "Since recently assuming command": Zumwalt, p. 187.
263 "The voice call *'Hellcat'*": Ibid., p. 189.
264 Kyocera has 2,000 employees: Lad Kuzela, "Putting Japanese-Style Management to Work," *Industry Week*, Sept. 1, 1980, p. 61.
264 Peter Vaill is a student: Peter B. Vaill, "Toward a Behavioral Description of High-Performing Systems," in *Leadership: Where Else Can We Go?*, ed. Morgan W. McCall, Jr., and Michael M. Lombardo (Durham, N.C.: Duke University Press, 1978), pp. 109–11.
265 "Caterpillar has always started": "Caterpillar: Sticking to Basics to Stay Competitive," *Business Week*, May 4, 1981, p. 76.
266 in one old plant: Edward Meadows, "How Three Companies Increased Their Productivity," *Fortune*, Mar. 10, 1980, p. 97.
267 "Bringing financial information": Charles G. Burck, "What Happens When Workers Manage Themselves," *Fortune*, July 27, 1981, p. 68.

Page

267   "Nothing is worse for morale": Richard T. Pascale, "The Role of the Chief Executive in the Implementation of Corporate Policy: A Conceptual Framework," Research Paper no. 357 (Graduate School of Business, Stanford University, February 1977), p. 39.

268   "A man wouldn't sell": Manchester, *Good-bye, Darkness,* p. 200.

269   At Mars, Inc.: Robert Levy, "Legends of Business," *Dun's Review,* June 1980, p. 92.

269   "Most Japanese corporations": Ohmae, p. 27.

270   *"Peter Peterson":* Ogilvy, "The Creative Chef," p. 209.

271   "What I am really talking about": Barry F. Sullivan, "International Service Products: The Opportunity of the 80s" (speech to the American Bankers Association, International Banking Symposium, Washington, D.C., Mar. 29, 1981), p. 13.

271   "I'm a big advocate": John S. McClenahen, "Moving GTE Off Hold," *Industry Week,* Jan. 12, 1981, p. 67.

271   "As divisions reach a certain size": Barron, "British 3M's Multiple Management," p. 54.

272   "Essentially, we act": Bro Uttal, "The Gentlemen and the Upstarts," p. 100.

273   "We don't need a 5,000 person plant": *Dun's Review,* December 1977, pp. 54–55.

274   "In the last decade": Roger L. Cason, "The Right Size: An Organizational Dilemma," *Management Review,* April 1978, p. 27.

274   "Skinner cites an episode": Lubar, p. 55.

274   The theoretical case is stated: John Child, *Organization: A Guide to Problems and Practice* (New York: Harper & Row, 1977), pp. 222–23.

276   "Teams set their own": Shepherd, "Innovation at Texas Instruments," p. 84.

277   "People can be themselves": E. F. Schumacher, *Small Is Beautiful: Economics as if People Mattered* (New York: Harper & Row, 1973), p. 75.

277   "One reason why the Roman Empire": Anthony Jay, *Management and Machiavelli: An Inquiry into the Politics of Corporate Life* (New York: Holt, Rinehart and Winston, 1967), pp. 63–64.

278   "Substituting rules for judgment": "The Iconoclast Who Made Visa No. 1." *Business Week,* Dec. 22, 1980, p. 44.

CHAPTER 9: HANDS-ON, VALUE-DRIVEN

Page

280   "Most contemporary writers": John W. Gardner, *Morale* (New York: Norton, 1978), p. 28.

Page

280 "Tough-minded managers": Julien R. Phillips and Allan A. Kennedy, "Shaping and Managing Shared Values," *McKinsey Staff Paper*, December 1980, p. 1.

280 "One may speculate": Watson, *A Business and Its Beliefs*, pp. 4–6.

281 "The formation of an institution": Selznick, *Leadership in Administration*, p. 28.

282 "The task of the leader": Hugh Sidey, "Majesty, Poetry and Power," *Time*, Oct. 20, 1980, p. 39.

282 "To create an institution": Selznick, pp. 151–53.

283 "There is a special relationship": *This Is Delta* (Atlanta: Delta Air Lines, 1981), p. 8.

283 "The Dana style": *Breaking with Tradition: Dana 1981 Annual Report* (Toledo, Ohio: Dana Corporation, 1981), p. 6.

283 "Availability of parts": *Caterpillar Annual Report 1981* (Peoria, Ill.: Caterpillar Tractor Co., 1981), p. 14.

283 "Digital believes": *Digital Equipment Corporation Annual Report 1981* (Maynard, Mass.: Digital Equipment Corporation, 1981), p. 12.

283 "Back in 1890": *Serving Customers Worldwide: Johnson & Johnson 1980 Annual Report* (New Brunswick, N.J.: Johnson & Johnson, 1980), p. 20.

284 "getting the best from the man": Kathleen K. Wiegner, "Corporate Samurai," *Forbes*, Oct. 13, 1980, p. 172.

284 "The cardinal responsibility": James MacGregor Burns, *Leadership* (New York: Harper & Row, 1978), p. 237.

285 James Brian Quinn believes: James Brian Quinn, "Strategic Goals: Process and Politics," *Sloan Management Review*, fall 1977, p. 26.

285 "I want all our people": David Ogilvy, *Principles of Management* (New York: Ogilvy & Mather, 1968), p. 2.

285 "Set and demand standards": Marshall Loeb, "A Guide to Taking Charge," *Time*, Feb. 25, 1980, p. 82.

285 "We want to give": Watson, p. 29.

285 "We believe that an organization": Ibid, p. 34.

286 "We have learned from experience": A. E. Pearson, *A Look at Pepsi-Co's Future* (Purchase, N.Y.: PepsiCo, December 1980), p. 10.

287 "Success in instilling values": Phillips and Kennedy, p. 8.

288 Harry Gray of United Technologies: "What Makes Harry Gray Run?" *Business Week*, Dec. 10, 1979, p. 77.

288 "I show up in places": Ibid., p. 80.

289 "Once a division": Hewlett and Packard, *The HP Way*, p. 10.

289 "Do not summon people": Ogilvy, *Principles of Management*, p. 2.

Page
289   "I travelled about 200,000 miles": Dowling "Conversation with Edward Carlson,"pp. 52–54.
290   "An important principle": Selznick, p. 110.
290   When he started those 200,000-mile years: Pascale, "The Role of the Chief Executive," pp. 37ff.
291   "Perhaps the most subtle challenge": Pearson, p. 3.
291   "You can't accomplish anything": Loeb, p. 82.
291   "Try to make working": Ogilvy, *Principles,* p. 2.

CHAPTER 10: STICK TO THE KNITTING

Page
292   "Back in the sixties": Young, "Views on Management," p. 3.
292   "In the wine and liquor business": Thomas J. Peters, "Structure as a Reorganizing Device: Shifting Attention and Altering the Flow of Biases," unpublished manuscript (September 1979), p. 34.
294   the first systematic study: Michael Gort, *Diversification and Integration in American Industry: A Study by the National Bureau of Economic Research* (Princeton, N.J.: Princeton University Press, 1962).
294   The most comprehensive study: Richard P. Rumelt, *Strategy, Structure and Economic Performance* (Graduate School of Business Administration, Harvard University, 1974).
294   "These companies have strategies": Ibid., p. 123.
295   To take a couple of example: Ibid., pp. 88–122.
295   study published in the *Journal of Finance:* Robert Haugen and Terence Langetieg, "An Empirical Test for Synergism in Merger," *Journal of Finance,* September 1975, pp. 1003–14.
296   One final study: Christopher Lorenz, "Pioneers: The Anti-Merger Specialists," *Financial Times,* Oct. 30, 1981, p. 16.
299   "Never acquire any businesses": "The Ten Best-Managed Companies," p. 30.
299   "This company has never": "P&G's New New-Product Onslaught," *Business Week,* Oct. 1, 1979, p. 79.
300   "Observers . . . say Boeing's strength": Victor F. Zonana, "Boeing's Sale to Delta Gives It Big Advantage over U.S. Competitors," *Wall Street Journal,* Nov. 13, 1980, p. 1.
300   "For years Deere": Bob Tamarkin, "The Country Slicker," *Forbes,* Jan. 21, 1980, p. 40.
300   "The wisdom guiding": Thomas Petzinger, Jr., "Indiana Standard Continues Its Strategy for Growth, Bucking the Takeover Trend," *Wall Street Journal,* Dec. 14, 1981, p. 12.

Page

303   "Imagine McDonald's": Gordon Weil, *Sears, Roebuck, U.S.A.: The Great American Store and How It Grew* (New York: Stein and Day, 1977), p. 255.

303   "Keith Crane's predecessor": Gail Bronson, "Colgate Works Hard to Become the Firm It Was a Decade Ago," *Wall Street Journal,* Nov. 23, 1981, pp. 1, 8.

305   "One can say that turbines": Sandra Salmans, "Demerging Britain's G.E.," *New York Times,* July 6, 1980, p. F7.

305   *Forbes* noted: Thomas Jaffe, "Is This It?" *Forbes,* Feb. 2, 1981, p. 48.

305   Litton's sell-off: Nick Galluccio, "The Housecleaning Is Over," *Forbes,* Nov. 24, 1980, p. 74.

CHAPTER 11: SIMPLE FORM, LEAN STAFF

Page

309   A wonderful example of simplicity: Nancy Kaible, "Johnson & Johnson," unpublished manuscript (San Francisco, Calif.: McKinsey & Co., November 1981).

309   "[J&J's] central staff": Ross A. Webber, "Staying Organized," *Wharton Magazine,* spring 1979, p. 22.

309   "We have periodically studied": "The 88 Ventures of Johnson & Johnson," *Forbes,* June 1, 1972, p. 24.

312   "The key is to get": Lynda Schuster, "Wal-Mart Chief's Enthusiastic Approach Infects Employees, Keeps Retailer Growing," *Wall Street Journal,* Apr. 20, 1981, p. 21.

313   Ford over the last twenty-four months: "A New Target: Reducing Staff and Levels," *Business Week,* Dec. 21, 1981, p. 69.

Index

Abernathy, William, 36–37, 38, 47, 197–199
Acheson, Dean, 47, 291
Acme Cleveland, 312
action principle, 13–14, 17, 119–155, 320
  communication in, 121–123
  environmental support for, 145–150
  experimentation as, 134–154
  flexibility in, 121–125
  learning process in, 143–145
  numbers in, 141–143
  orientation for, 154–155
  project teams in, 131–134
  small groups in, 125–127
  system simplification as, 150–154
  task forces in, 127–132
  worker motivation as, 123–124
Activision, 262
  prototype importance at, 136
Adams, Robert M., 137, 231
adhocracy, 127–131
  bureaucracy vs., 121, 134, 314
Administrative Behavior (Simon), 101
Age of Discontinuity (Drucker), 111
Allen, Thomas, 220
Allen, Woody, 119
Allen-Bradley, 188–189
  cooperative efforts of, 195
AM International, 46
Ambiguity and Choice in Organizations (March and Olsen), 102
American Airlines, 191
American Challenge (Servan-Schreiber), 33–34

Ames, B. Charles, 151, 152, 312
Amoco:
  acquisition strategy of, 300
  drilling success of, 141, 193, 210–211
Anderson, David, xiii, 111, 212
Anderson, Richard, 176
Andrews, Kenneth, 97
Ansoff, Igor, 111
Apple, 141, 286–287
Arco, 193
Argyris, Chris, 49
Art of Japanese Management, The (Athos and Pascale), 11
Ash, Roy, 45–46
Athos, Anthony, 9, 11, 101
  on good managers, 29
Atlantic Monthly, 34
AT&T, 80
authority, acceptance of, 78–80
auto industry, 109, 252
  overextension in, 112
  U.S. vs. Japanese, 34, 37
autonomy and entrepreneurship, 14, 52–54, 200–234
  at Dana, 112, 249
  through discipline, 322
  "godfather" in, 208–209
  internal competition and, 215–218
  "pillar" for, 315–317
  "simulated," 212–214
  in 3M, 224–234
  see also championship; chunks, theory of; communication
Avon, 190
Aweida, Jesse, 142–143

☗ HarperBusiness ESSENTIALS

Business Classics for the Inquisitive Executive

THE INNOVATOR'S DILEMMA
Clayton M. Christensen
A *NEW YORK TIMES*, *WALL STREET JOURNAL*, AND *BUSINESS WEEK* BESTSELLER
ISBN 0-06-052199-6 (paperback)

"The book to read among mainstream managers trying to dope out an Internet strategy."
— *New York Times*

THE INTELLIGENT INVESTOR: *Revised Edition*
Benjamin Graham
ISBN 0-06-055566-1 (paperback)

First published in 1949, this is the definitive book on investing due to Graham's timeless philosophy of "value investing." *Now with new commentary by Jason Zweig.*

THE WISDOM OF TEAMS: *Creating the High-Performance Organization*
Jon R. Katzenbach & Douglas K. Smith
ISBN 0-06-052200-3 (paperback) • ISBN 1-55994-967-8 (audio)

"A thoughtful and well-written book filled with fascinating examples. . . .
You'll be hard-pressed to find a better guide." —John Byrne, *Business Week*

INDECENT EXPOSURE: *A True Story of Hollywood and Wall Street*
David McClintick
With a Foreword by James B. Stewart and a new Afterword by the author
A NATIONAL BOOK AWARD FINALIST AND NATIONAL BESTSELLER
ISBN 0-06-050815-9 (paperback)

"Fascinating. . . . One of the best-reported and best-written business adventures
ever published." —*Business Week*

BUILT TO LAST: *Successful Habits of Visionary Companies*
James C. Collins & Jerry I. Porras
ISBN 0-06-051640-2 (paperback) • ISBN 0-694-51479-9 (audio)

"One of the most eye-opening business studies since *In Search of Excellence*."
—Kevin Maney, *USA Today*

THE EFFECTIVE EXECUTIVE
Peter F. Drucker
ISBN 0-06-051607-0 (paperback)

"An imaginative book. . . . A survival manual on how to escape organization traps."
—*Wall Street Journal*

▲ HarperBusiness ESSENTIALS

Business Classics for the Inquisitive Executive

CROSSING THE CHASM
Marketing and Selling High-Tech Products to Mainstream Customers
Geoffrey A. Moore
ISBN 0-06-051712-3 (paperback) • PerfectBound e-book

"A must-read for marketing executives, CEOs, and especially venture capitalists."
—Jeff Miller, President, Documentum

THE LEADERSHIP ENGINE
How Winning Companies Build Leaders at Every Level
Noel M. Tichy with Eli Cohen
A *WALL STREET JOURNAL* BESTSELLER
ISBN 0-88730-931-3 (paperback) • ISBN 0-694-51881-6 (audio) • PerfectBound e-book

Practical advice from the acclaimed authority on organizational transformation.

REENGINEERING THE CORPORATION
A Manifesto for Business Revolution
Michael Hammer and James Champy
ISBN 0-06-055953-5 (paperback) • PerfectBound e-book

"May well be the best-written, most well-reasoned business book for the managerial masses since *In Search of Excellence.*" —*BusinessWeek*

LEADERS: *Strategies for Taking Charge*
Warren G. Bennis & Burt Nanus
ISBN 0-06-055954-3 (paperback)

"An insightful book that should be read by every manager aspiring to be a true leader. . . . One of the most important books of its type in a long time." —*Chicago Tribune*

BARBARIANS AT THE GATE: *The Fall of RJR Nabisco*
Bryan Burrough and John Helyar
ISBN 0-06-053635-7 (paperback)

"The fascinating inside story of the largest corporate takeover in American history. . . . It reads like a novel." —*Today Show*

Don't miss the next book by your favorite author.
Sign up for AuthorTracker by visiting *www.AuthorTracker.com.*

Available wherever books are sold, or call 1-800-331-3761 to order.